Astronomers' Observing Gui

Other titles in this series

Double and Multiple Stars and How to Observe Them
James Mullaney

The Moon and How to Observe It
Steven R. Coe

Related titles

Field Guide to the Deep Sky Objects
Mike Inglis

Deep Sky Observing
Steven R. Coe

The Deep-Sky Observer's Year
Grant Privett and Paul Parsons

The Practical Astronomer's Deep-Sky Companion
Jess K. Gilmour

Observing the Caldwell Objects
David Ratledge

Choosing and Using a Schmitt-Cassegrain Telescope
Rod Mollise

Julius Benton

Saturn and
How to Observe It

With 96 Figures, 86 in Full Color

 Springer

Dr. Julius L. Benton, Jr. BS, MS, PhD.
Fellow of the Royal Astronomical Society
Association Lunar and Planetary Observers (ALPO) Saturn Section
Wilmington Island
Savannah, 6A 31410
USA
Jlbaina@msn.com

Series Editor

Dr. Mike Inglis BSc, MSc, PhD.
Fellow of the Royal Astronomical Society
Suffolk County Community College, New York, USA
inglism@sunysuffolk.edu

British Library Cataloguing in Publication Data
A catalog record for this book is available from the British Library

Library of Congress Control Number: 2005925511

Astronomers' Observing Guides Series ISSN 1611-7360
ISBN-10: 1-85233-887-3   e-ISBN 1-84628-045-1
ISBN-13: 978-1-85233-887-3
Springer Science+Business Media
springeronline.com

Typeset by EXPO Holdings Sdn Bhd
Printed in Singapore
58/3830-543210   Printed on acid-free paper

9 8 7 6 5 4 3 2 1

This book is dedicated to my father and mother, Julius and Susan Benton, who gave me my first telescope, to my aunt, Mary Ann Jones, who always encouraged my explorations of the heavens, to my family for their patience and understanding as my astronomical endeavors consumed countless hours all these years, to my mentor and longtime friend Walter Haas, and with deepest gratitude to Saturn observers everywhere who faithfully contributed observations to the A.L.P.O. Saturn Section during my tenure as Section Coordinator.

Contents

Introduction 1

1. Saturn as a Planet 5

2. Telescopes and Accessories 51

3. Factors that Affect Observations 75

4. Visual Impressions of Saturn's Globe and Ring System 89

5. Drawing Saturn's Globe and Rings 111

6. Methods of Visual Photometry and Colorimetry 123

7. Determining Latitudes and Timing Central Meridian Transits 133

8. Observing Saturn's Satellites 141

9. A Primer on Imaging Saturn and Its Ring System 147

Appendix A: Association of Lunar and Planetary Observers (ALPO)
   Saturn Section Observing Forms 164

Appendix B: Bibliography 174

Index 179

Introduction

I received my first telescope, a 60 mm (2.4 in) Unitron refractor, as a surprise Christmas gift from my father when I was 10 years old, and over the next several years, I spent countless hours exploring the heavens, seeking out virtually every celestial object I could find with this small aperture. I consider myself quite fortunate to have been blessed with a dark, unobstructed observing site for most of my childhood, unlike many of my astronomical friends who were always trying to get to a remote location away from city lights to do worthwhile deep-sky observing. I only had to carry my telescope and star charts just a few feet away into my backyard.

By the time I entered high school, the night sky had become a delightfully familiar place. I had tracked down virtually all of the galaxies, nebulae, and star clusters within reach of my little instrument, and I split most of the double stars that were theoretically possible with its exquisite optics. Eventually, I earned sufficient funds working part-time jobs (and saving school lunch money) to purchase a premium 10.2 cm (4.0 in) refractor, another Unitron that I quickly put through its paces, once again surveying my favorite deep-sky objects. Despite the fact that I could see all of them much better with increased aperture, I soon recognized how virtually changeless they were, so I started expanding my observational pursuits. Taking advantage of the increased resolution of the new refractor, my interests evolved almost exclusively to observing the moon and the brighter planets. The wealth of detail I could see on the moon, Jupiter, and Saturn at 250× thoroughly fascinated me, and countless evenings were occupied watching variations in their appearance. Undeniably, why I ended up being chiefly a planetary observer had more to do with the continually changing aspect of the moon and planets than anything else. Now that I had substantially improved aperture at my disposal, as well as a growing collection of eyepieces and accessories, I naturally wanted to observe the moon and planets more often. I soon realized, however, that one of the drawbacks of a larger instrument is decreased portability. Yet, having to carry my clock-driven telescope from place to place never became much of an issue, except when I wanted to view comets or asteroids, for experience had taught me that lunar and planetary observing did not necessarily require dark skies or even an absolutely clear horizon. Like I had done so many times in the past, all I needed to do was step right outside my door to observe. And so, I routinely followed the moon and most of the brighter planets throughout many wonderful observing seasons. Saturn, though, ultimately became my favorite solar system object, and it was not long before I adopted the practice of making careful drawings and writing down all of my observational notes in a logbook for future reference.

My interest in lunar and planetary astronomy followed me well into my college years. The rather gloomy outlook at the time for potential employment opportunities in astronomy forced me to select a major in physics and the environmental sciences, but I always made sure I was never far away from my telescopes on clear nights! By the time I completed my undergraduate studies, my interest in lunar and planetary astronomy had become a virtual obsession, partially ushered in by the unprecedented events of July 20, 1969, when Apollo 11 touched down on the moon in the Sea of Tranquility. The following year I attended my first astronomical convention, where for the first time I met Walter Haas, the founder and then executive director of the Association of Lunar and Planetary Observers (ALPO). I discussed with him my deep interest in contributing worthwhile lunar and planetary observations, and his enthusiasm, encouragement, and guidance helped stimulate my involvement in many ALPO activities. For a few more years, however, the rigors of graduate school occupied much of my available time as I pursued advanced degrees, but I still somehow managed to set aside a few hours a week to spend time at my telescopes recording observations systematically.

The congenial, informal atmosphere of the ALPO proved to be refreshing and wonderfully captivating, and I soon developed a real appreciation for the great diversity of backgrounds and experience of the people I met and corresponded with. Sharing many different philosophical and scientific viewpoints about instrumentation and observing proved to be very meaningful over the years. Although I often found serious observing endeavors to be challenging work, they were also enormous fun. I enthusiastically welcomed the opportunity to contribute all of my own observations to a collective pool of data that had the potential for enhancing our knowledge about Saturn and the solar system as a whole. Through an active exchange of information and ideas in a collective forum, many individuals improved as observers, including myself, and some even went on to become professional astronomers. In addition, I had not been a member of the ALPO for very long before I discovered that the *The Strolling Astronomer* (also known as the *Journal of the Association of Lunar and Planetary Observers*) was essential reading. Most of the information contained within its pages seldom existed elsewhere, and this publication for many years has helped establish and preserve a vital link between the amateur and professional astronomical communities that might not otherwise exist. Annual conventions, often held jointly with other national and international groups, were always enjoyable events as much as they proved to be intellectually stimulating. I can attribute many lasting friendships to such meetings.

In 1971 I was appointed coordinator of the ALPO Saturn Section. I was truly honored to be selected to serve in such a role, and I valued the confidence placed in me by my mentor, Walter Haas, and other ALPO colleagues. Any small contribution that I have been able to make to what we know about the planet Saturn from the standpoint of observational astronomy has come as a sincere labor of love, something I have never grown tired of even after nearly 34 years of recording, analyzing, and publishing detailed apparition reports. But whatever success the ALPO Saturn Section has achieved, none of it would have been possible without the enthusiasm and perseverance of many dedicated observers too numerous to mention here.

Like many of my contemporaries, I consider myself very fortunate to have grown up during the Space Age, witnessing firsthand the enormous revelations and progress made in planetary science. It has been fascinating to watch the marvelous transformation of our nearest celestial neighbors from virtually unknown and inaccessible objects into much more familiar worlds over little more than three decades. With such rapidly occurring advances, I suppose it may be tempting to conclude that the work of amateur astronomers long ago passed into obsolescence from our fixed and limited vantage point in space. And, yes, it is obvious that lunar explorations by the Apollo astronauts or the close surveillance of planets and satellites by orbiting, impacting, or roving spacecraft are clearly beyond the domain of the Earth-based amateur astronomer. But make no mistake about it: there are still many areas of lunar and planetary observation where the work of amateur astronomers has not been outmoded by the onslaught of prohibitively expensive and imposing equipment. Unlike many of their professional counterparts, amateur astronomers continue to enjoy the virtual freedom and advantage of being able to study their favorite solar system objects for extended periods of time and precisely when they want to. Indeed, the greatest potential visual observers have for making useful contributions to science is a systematic, long-term, and simultaneous monitoring of the moon and planets at wavelengths of light to which the eye has greatest sensitivity. An enduring advantage that trained eyes of skilled amateurs have is the unique ability to perceive, at intermittent moments of exceptional seeing, subtle detail on the surfaces and in the atmospheres of solar system bodies that frequently escapes normal photography with considerably larger apertures. And while being careful not to abandon fundamentally important systematic visual work, more and more observers are employing sophisticated electronic devices such as charged couple devices (CCDs), specialized video cameras, and webcams to record impressive, detailed images of the planet, far surpassing what had been previously possible by astrophotography. Furthermore, well-organized systematic work by dedicated amateurs has increasingly caught the attention of the professional community, evidenced by several invitations extended to them for participation in specialized research projects. Simultaneous observing programs and close professional–amateur alliances have already carried over into the 21st century and will undoubtedly grow in the coming years.

If we considered the planet only as a globe, Saturn would be a somewhat smaller, dimmer, and relatively quiescent replica of the giant Jupiter. But, with its broad symmetrical rings, Saturn is an object of exquisite and unsurpassed beauty, holding a particular magnetism for the visual and photographic observer alike. Aside from its obvious aesthetic qualities, the planet exhibits numerous features requiring persistent and meticulous observation, plus eight satellites that are readily accessible to moderate-size telescopes if observers know where to look. In this book, the reader will learn about how to observe Saturn, its rings, and brighter moons using methods, techniques, instruments, and accessories that are readily available to amateur astronomers. One of the major objectives I hope to accomplish is to first acquaint the reader with some fundamental, up-to-date information about Saturn as a planet, then focus on the basics involved in recording useful data and reporting observational results, plus offer suggestions for more advanced and specialized work. Observers will discover how they can take part in well-organized Saturn research programs conducted by international organizations such as the ALPO or the British Astronomical Association (BAA), which share data with the professional community regularly and publish detailed observational reports. As readers utilize the methods and techniques described in this book, the need for even more comprehensive information on certain endeavors will undoubtedly arise. Accordingly, I will always be delighted to correspond with interested parties and provide guidance and advice, including recommendations for more in-depth observational pursuits. For added convenience, the two Internet links—ALPO official Web site: http://www.lpl.arizona.edu/alpo; and ALPO eGroup: Saturn-ALPO@yahoogroups.com—are international sites that readers can visit to access (and download) observing forms and instructions, ephemerides, special alerts and bulletins, results of recent observations, and a wealth of other timely information about Saturn, including ways to interface with professional astronomers. These sites also have abundant sub-links to a host of other important professional and amateur web sites of interest to the Saturn enthusiast, and observers can participate as they desire in discussion forums and promptly exchange information with one another. In addition to the above web sites, I have provided at the end of this book a fairly comprehensive list of references, including well-known astronomical periodicals and authoritative texts that should help readers learn more about the history of amateur observations, as well as the latest developments in planetary science, in particular our rapidly growing knowledge about Saturn.

Julius L. Benton, Jr.
Coordinator ALPO Saturn Section
Association of Lunar and Planetary Observers
c/o Associates in Astronomy
305 Surrey Road
Wilmington Park
Savannah, GA 31410
E-mail: jlbaina@msn.com

Saturn as a Planet

A Simplified View of the Solar System

In a hypothetical spacecraft looking down from far above our solar system, the luminosity of the sun would totally overwhelm that of the nine planets, all of which shine mostly by reflected light. Even the giant Jupiter would be hopelessly immersed in the solar glare. Consider also the fact that the mass of the sun is over 99.8% of the total mass of the known solar system—in contrast with our own geocentric perspective—which essentially relegates the planets, including Earth, to little more than orbiting debris! Yet, it is notable that the orbital motion of the planets comprises nearly 99% of the angular momentum of the solar system. The sun and planets are quite different, too. The sun is mostly plasma generated by nuclear fusion, while the planets are fundamentally solid rocky bodies composed of silicates, metals, ices, as well as varying amounts of liquid or gaseous constituents.

Of the nine planets that compose our solar system (Fig. 1.1), Mercury and Venus move around the sun in smaller orbits than that of the Earth—thus their classification as inferior (or sometimes "interior") planets. The rest of the planets revolve about the sun along orbits that are external to the Earth's orbit and are referred to as superior planets. In another nomenclature scheme, Mercury, Venus, Earth, and Mars are designated terrestrial planets because of their compositional similarity and relatively slow rotation. They are all mostly rocky and metallic bodies with fairly high bulk density, sporting diameters of 4878 km (Mercury), 12,100 km (Venus), 12,800 km (Earth), and 6878 km (Mars). Terrestrial planetary surfaces exhibit varying numbers of craters caused by meteoritic or cometary impacts since the solar system formed 4.6×10^9 years ago, as well as evidence of tectonic activity, such as faulting and volcanism. Their atmospheres are either for the most part nonexistent or are comparatively thin, made up of variable concentrations of gases like carbon dioxide (CO_2), nitrogen (N_2), and oxygen (O_2). Only two of the terrestrial planets have satellites, Earth and Mars, although our singular moon, with a diameter of 3474 km, is considerably more significant than the diminutive two rocky bodies, Phobos and Deimos, orbiting Mars. The Earth, of course, is unique with its ubiquitous life forms and oceans of liquid H_2O.

The giant planets Jupiter, Saturn, Uranus, and Neptune, with diameters of 143,000 km, 120,600 km, 51,100 km, and 49,500 km, respectively, belong to a different class known as the Jovian planets. They have strong magnetic fields, rotate rapidly, and their compositions are dominated by 75% to 90% hydrogen (H_2) and 10% to 25% helium (He), with varying amounts of water (H_2O), ammonia (NH_3), methane (CH_4), and other trace substances. All four Jovian planets have a rather

Figure 1.1. In this solar system montage of recent spacecraft images, the terrestrial planets (Mercury, Venus, Earth and its moon, and Mars) are roughly to scale with each other; likewise, the Jovian planets (Jupiter, Saturn, Uranus, and Neptune) are nearly to scale. Pluto does not appear in this assemblage of images. North is at the top in this image. (Credit: NASA and Jet Propulsion Laboratory, Pasadena, California.)

large number of accompanying satellites ranging in size from tiny moonlets, barely a few kilometers across, to exotic worlds, many sporting impact craters of all sizes, ice fields and wrinkled terrain, active volcanoes, and a host of other unique characteristics. A few are even larger than Mercury and our moon, and at least one, Saturn's Titan, has a fairly substantial atmosphere. Another distinguishing attribute of the giant planets is the presence of rings, most of them made up of rocky or icy debris, but none of the other Jovian planets has rings that come even close to rivaling the broad, majestic system that encircles Saturn.

Lastly, there is Pluto with a diameter of 2274 km, which presumably is composed of 70% rocky material and 30% icy substances, accompanied by a satellite Charon that is about half its size. There is a continuing dispute among astronomers as to whether Pluto should truly be considered a planet or demoted and thought of only as one of the larger asteroids or cometary bodies.

In addition to the nine planets and their natural satellites, there are two classes of smaller bodies orbiting the sun, the asteroids and comets, which represent material left over following the formation of the solar system. Most asteroids exist in orbits between Mars and Jupiter, but may also occupy gravitationally stable regions 60° ahead of and 60° behind major planets like Jupiter (e.g., the Trojan asteroids). Asteroids have diameters less than 1000 km, are composed mostly of rocky or metallic substances, and most have roughly circular orbits that generally lie within several degrees of the plane of the solar system. Comets are icy bodies a few kilometers across that can produce gaseous haloes and long wispy tails of considerable size and brightness when they approach the vicinity of the sun. Most comets are situated in the Oört cloud well beyond the major planets, but a few orbit near Neptune in the Kuiper belt, and they can have orbits of varying eccentricity and inclination. When they are far from the sun and very faint, it is sometimes extremely difficult to distinguish a comet from an asteroid.

Saturn: Basic Characteristics and Terminology

With a diameter of 120,600 km, Saturn resides at a mean distance from the sun of 9.54 astronomical units (AU), where 1 AU is equal to 1.43×10^9 km (the Earth's mean heliocentric distance), and completes one orbit in 29.5^y. It has a mean synodic period of 378^d (the time elapsed between one conjunction of Saturn with the sun to the next), so one apparition of the planet lasts slightly longer than one terrestrial year. Saturn's annual eastward motion relative to the background stars is approximately 12°; thus, it can remain in one constellation for quite some time. From perihelion to aphelion, Saturn's distance from the sun varies from 9.01 to 10.07 AU, respectively, which translates into a moderate orbital eccentricity of 0.056. Saturn's orbit is inclined to the ecliptic by 2.5°.

At opposition, Saturn can approach the Earth as close as ~8.0 AU and attain a maximum brightness of −0.3 m_v (where m_v denotes visual magnitude), and it can outshine every star in the sky except Sirius and Canopus. Even so, because of its comparatively large distance from the sun, it is considerably fainter than Jupiter or Mars (when Mars is near opposition), but this is not the only factor that affects the brilliance of Saturn as seen by observers on Earth. Because the majestic ring system is so highly reflective, the brightness of the planet as a whole is substantially controlled by the orientation of the rings to our line of sight. So, at opposition, and when the rings are open to their maximum extent, Saturn is at its brightest. When the rings appear edge-on to us, however, the visual magnitude of the planet may never surpass +0.8 m_v at opposition. Saturn has a Bond albedo, which is the total percentage of sunlight reflected by a planet in all directions, of 0.33. It

Figure 1.2. Saturn displays its familiar banded structure, with haze and pastel colors of NH_3–CH_4 (ammonia–methane) clouds at various altitudes in this Hubble space telescope (HST) image taken on March 22, 2004. The magnificent rings, seen here near their maximum tilt of 26.7° toward Earth, exhibit subtle hues suggesting chemical differences in their icy composition. North is toward the top in this image. (Credit: NASA, European Space Agency, and Erich Karkoschka, University of Arizona.)

has a visual geometric albedo, p, or percentage of sunlight reflected at 0° phase angle (full phase), of 0.47.

The tilt angle between Saturn's axis of rotation and the pole of its orbit, or obliquity, is 26.7° (the Earth has the familiar value of 23.5°). Although the planet's rotational axis maintains roughly the same orientation in space, it is obliquity that causes one hemisphere of Saturn to be tipped toward or away from the sun at certain points in its orbit. So, just like the Earth, Saturn exhibits seasons.

In almost any telescope, Saturn is a magnificent, enchanting object, but this is chiefly because of its best-known feature, the exquisite ring system (Fig. 1.2). In the eyepiece of a moderate-size telescope, the planet has a distinctly yellowish color, and at opposition (when Saturn is opposite the sun in the sky and can be seen virtually all night) the globe may reach a maximum angular equatorial diameter of 19.5″. As Saturn orbits the sun (as seen from the Earth), the most dramatic consequence of the Saturnian seasons is the varying presentation of the rings. Because the ring system is precisely in Saturn's equatorial plane, when the planet's North or South Pole is tipped toward the sun, summer occurs in that hemisphere. During Saturnian summer, the rings reach their maximum opening of 26.7° to the sun and to our vantage point on Earth, and they are at their brightest. Winter occurs in the opposite hemisphere of Saturn, which is tipped away from the sun and Earth, and most of that region is typically hidden from view by the rings as they pass in front of the globe. It is during spring and fall on Saturn that the rings can be oriented edgewise to the sun and to our line of sight, sometimes even disappearing for a period of time, even in a large telescope. These are periods during the Saturnian year when equal portions of the Northern and Southern Hemispheres of the planet become visible in our telescopes. The oblateness of Saturn's globe, or deviation from a perfect sphere because of rotation, is immediately obvious and amounts to 0.108.

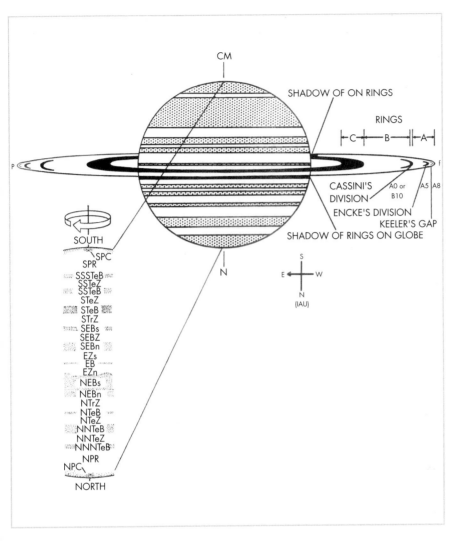

Figure 1.3. Diagram of the major features of Saturn's globe and rings, South is toward the top and east is to the left in this normally inverted view as seen in most astronomical telescopes, and features move across the globe from right to left (west to east) in the International Astronomical Union (IAU) convention. (Credit: Julius L. Benton, Jr., ALPO Saturn Section.)

Similar to Jupiter, Saturn displays a series of dark-yellowish to tan cloud belts and white to pale-yellowish zones extending across the globe roughly parallel to the equator and plane of the ring system. On occasion, discrete detail is visible in the belts and zones on the globe of Saturn. Any features seen on the planet may be similar in form to those on Jupiter, but a greater depth of overlying haze makes them poorly visible most of the time. Projections or appendages from belts, sometimes leading into extended dusky festoons, or bright spots in the zones, comprise the most frequently recorded types of phenomena. Figure 1.3 is a diagram of Saturn's globe and rings (as seen in a typical inverting astronomical telescope) with the major atmospheric features and ring components that are sometimes

Table 1.1. Nomenclature and characteristics of major saturnian global features*

Feature	Description	Notes
SPR	South polar region	This is usually the most southerly part of the globe, sometimes differentiated into a SPC (south polar cap) in the extreme south. The SPR is usually quite variable in appearance, usually quite dusky but can be bright on occasion.
SSTeZ	South south temperate zone	This zone typically separates the SSTeB and the SPR; it is usually much duller than the other southern hemisphere zones.
SSTeB	South south temperate belt	Infrequently visible, but even when seen, it is a very thin, ill-defined feature.
STeZ	South temperate zone	This zone is usually quite bright, sometimes showing faint wispy dark features and occasional bright spots at the threshold of vision.
STeB	South temperate belt	This belt is usually visible, and it has occasionally displayed some poorly defined dark spots.
STrZ	South tropical zone	Like the STeZ, this zone is also quite bright, and it periodically may exhibit dusky phenomena and small, diffuse bright spots or regions.
SEB	South equatorial belt	This belt is often quite dark, easily seen, and usually is differentiated into the SEBs and SEBn (southern and northern components, respectively), separated by a brighter SEBZ (south equatorial belt zone). The SEB frequently displays more activity than do the other belts in the southern hemisphere of Saturn's globe.
EZ	Equatorial zone	Almost without exception, the EZ is the brightest zone on the planet's globe, and dusky details and white spots have been observed in this zone with a greater frequency than in other global regions. The very thin and rarely seen EB (equatorial belt) separates the EZ into the EZs and EZn (southern and northern components, respectively).
NEB	North equatorial belt	Exhibits many of the same characteristics of its counterpart in the southern hemisphere of Saturn, the SEB, including differentiation into an NEBs, NEBn, and NEBZ.
NTrZ	North tropical zone	Like the STrZ, this feature lies between two dusky belts and is often quite bright, showing activity occasionally in the form of festoons or whitish mottlings.
NTeB	North temperate belt	Usually visible, this belt has shown activity from time to time in the form of dark spots or disturbances.
NTeZ	North temperate zone	A fairly bright zone, similar in characteristics and periodic activity to the STeZ.
NNTeB	North north temperate belt	Seldom reported, but under optimum circumstances, it may be barely perceptible as a delicate, linear belt crossing the globe.
NNTeZ	North north temperate zone	A somewhat dull zone sometimes seen separating the NNTeB and the NPR.

Table 1.1. Continued

Feature	Description	Notes
NPR	North polar region	This is the northernmost part of the globe, often quite dusky in its overall appearance, but sometimes it can brighten; at the extreme northern limb, a NPC (*north polar cap*) can occasionally be seen.

* Sequence follows that in *Fig. 1.3*; visibility of some global features is affected by the location and orientation of the ring system.

visible. Table 1.1 gives brief details on the nomenclature and some rough (not necessarily typical) characteristics of the main belts and zones of Saturn's globe.

The equatorial regions (NEB, SEB, and EZ; see Table 1.1) have a sidereal rotation period of $10^h14^m00^s$, and this region is designated system I. The remainder of the globe of Saturn is called system II with a sidereal rotation period of $10^h38^m25^s$, although the SPR and NPR are sometimes excluded and assigned a rotation rate that is equal to system I. A system III radio rate of $10^h39^m22^s$ has also been determined for the interior of Saturn. Latitudes of belts and zones on the globe are not appreciably altered by rotation, and as shown in Figure 1.3, features move across the planet from right to left (in the normal inverted view) in a West-to-East, or prograde, fashion. Note that west and east here corresponds to true directions on Saturn as adopted by the International Astronomical Union (IAU), which are opposite normal sky directions when viewing Saturn from Earth, and we will adopt this convention without exception throughout this book. In addition, readers should be aware that the preceding (*p*) limb of Saturn is to the east and the following (*f*) limb is to the west in the IAU sense in the normal inverted view of an astronomical telescope.

By terrestrial standards, Saturn is a giant planet, despite the fact that it is a little less than a third the mass of Jupiter. As determined by interactions of the planet with its family of natural satellites, the mass of Saturn has been determined to be 5.68×10^{26} kg. The mean density of 700 kg/m^3 for Saturn is the lowest of the major planets, and realizing that water (H$_2$O) has a density of 1000 kg/m^3, Saturn would float on a sufficiently large hypothetical ocean!

Even though Jupiter, Uranus, and Neptune are now among the family of planets known to have rings encircling them, those of Saturn are clearly in a class all by themselves. The Saturnian ring system is considerably brighter and more complex than any of the others throughout the solar system, with an albedo higher than that of the globe. So the rings contribution to the total brilliance of Saturn as a planet is substantial. Across their major axis, the rings have an angular extent of as much as 44.0″. The ring system lies in the equatorial plane of Saturn, so they are inclined by the same obliquity of 26.7°. As Saturn orbits the Sun in 29.5^y (with the planet's axis of rotation maintaining the same orientation in space) the intersection of Earth's orbit and the plane of the ring system occur twice, at intervals of roughly 13.75^y and 15.75^y. The two periods are of unequal duration because of the ellipticity of Saturn's orbit. The orbital intersection points signify times when the rings appear edgewise to our line of sight. In the shorter period, the southern face of the rings and southern hemisphere of the globe of the planet is inclined toward Earth, and Saturn passes through perihelion during this interval; the tilt of the rings, as

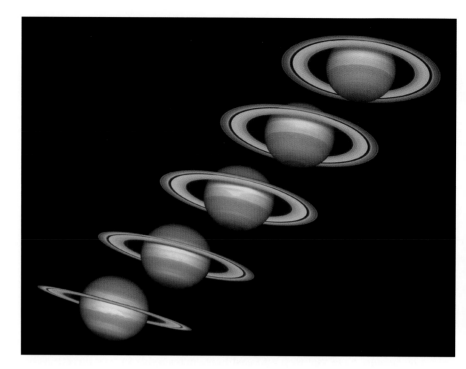

Figure 1.4. These Hubble space telescope images captured from 1996 to 2000 show Saturn's rings open up from just past edge-on to nearly fully open as it moves from autumn towards winter in its northern hemisphere. Saturn's equator is tilted relative to its orbit by 26.7°, very similar to the 23.5° tilt of the Earth. As Saturn moves along its orbit, first one hemisphere and then the other tilts toward the sun. This cyclical change causes seasons on Saturn, just as the changing orientation of Earth's tilt causes seasons on our planet. North is at the top. (Credit: NASA and Hubble space telescope [HST] Heritage Team [STScI/AURA].)

we see them on Earth (as shown in Fig. 1.4), varies from 0° to −26.7° and back to 0° again. In the longer period, Saturn passes through aphelion, and the north face of the rings and the northern hemisphere of the globe are exposed to observers on Earth, and the tilt of the rings to our line of sight changes from 0° to +26.7° and then back to 0°. Because the rings are no more than ~100 m thick, they become extremely hard to see, or they may seem to disappear entirely, when edge-on.

The rings lie inside Saturn's Roche limit, or within the minimum distance from the planet where a body with no appreciable gravitational cohesive forces can exist without disruption. The classical ring system is composed of three major ring components. The first is ring A, the usually seen outermost component with an inner radius from the center of Saturn of 122,200 km, an outer radius of 136,800 km, and a width of 14,600 km. Next is ring B, the central, broader, and brighter ring with an inner radius of 92,000 km, an outer radius of 117,500 km, and a width of 25,500 km. Last is ring C, the dusky inner Crape ring with an inner radius of 74,658 km, an outer radius of 92,000 km, and a width of 17,342 km. Taken together, the classical ring system from tip to tip extends 273,600 km. Rings A and B are separated by a dark gap roughly 4800 km wide, known as Cassini's division. It is visible in small telescopes with good optics and cooperative seeing conditions. About halfway between the inner and outer limits of ring A is Encke's complex with a width of

~320 km, but it is not as well defined as Cassini's gap. About 3200 km from the outer edge of ring A is Keeler's gap, with a width of about 35 km, and although it is visible from Earth, it requires very large apertures to be seen to advantage. All of the aforementioned major ring components and major divisions (including the faint Keeler's gap), which are usually observable from Earth, are illustrated in Figure 1.3.

Ring D, an extremely faint component internal to ring C, has an inner radius of ~67,000 km, an outer radius of 74,510 km, and a width of 7,510 km, and seems to extend downward nearly to the cloud tops of Saturn's atmosphere. Situated just outside ring A is a very narrow ring component, ring F, with an inner radius of 140,210 km, an outer radius of 140,600 km, and a slender width of no more than 390 km or so. Beyond ring F is the extremely tenuous ring G with a radius of 165,800 km, an outer radius of 173,800 km, and a width of 8000 km. Finally, there is the extraordinarily diffuse ring E with an inner radius of 180,000 km, an outer radius of 480,000 km, and a width of 300,000 km, and is the outermost ring component. The rings of Saturn will be discussed in greater detail later, including their interaction with some of the satellites. With perhaps the exception of ring E, it is generally held that these ring components are beyond the reach of the Earth-bound visual observer.

The Atmosphere of Saturn

The composition of Saturn's atmosphere is ~93% hydrogen (H_2) and ~5% helium (He). Minor constituents include methane (CH_4) and ammonia (NH_3), but the greatest percentage of the latter is in liquid or solid form in Saturn's extremely frigid upper atmosphere where temperatures are ~95 K. Traces of water vapor, ethane (C_2H_6), and other compounds occur as well (Table 1.2). Saturn has the lowest mean density of all of the planets at 700 kg/m³, another indication that the planet has a very H-rich interior with few rocky materials.

Although H_2 and He dominate in the atmosphere of Saturn, accounting for ~98.0% of the gaseous constituents, the amount of He is significantly less than that found in the atmosphere of Jupiter. During Saturn's early history, it is likely that the

Table 1.2. Main constituents of Saturn's atmosphere

Hydrogen	H_2	93%
Helium	He	5%
Methane	CH_4	0.2%
Water vapor	H_2O	0.1%
Ammonia	NH_3	0.02 %
Ethane	C_2H_6	0.0005%
Phosphine	PH_3	0.0001%
Hydrogen sulfide	H_2S	<0.0001%
Methylamine	CH_3NH_2	<0.0001%
Acetylene	C_2H_2	Trace
Hydrogen cyanide	HCN	Trace
Ethylene	C_3H_4	Trace
Carbon monoxide	CO	Trace

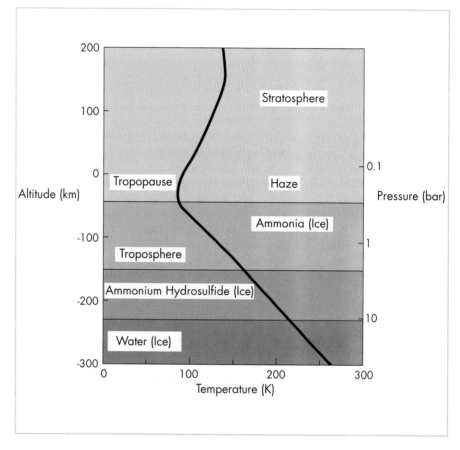

Figure 1.5. Simplified vertical structure of Saturn's atmosphere. (Credit: Julius L. Benton, Jr., ALPO Saturn Section.)

more massive He sank toward the center of the planet during the differentiation process, leaving the outer atmospheric layers rich in H_2 and relatively depleted in He. Figure 1.5 is a very simplified diagram of the vertical structure of Saturn's atmosphere.

Saturn's clouds are not sufficiently transparent to most wavelengths of light to permit detailed studies of the deeper levels of the planet's atmosphere. Examination of the planet's thermal radiation at infrared (IR) wavelengths shows that the temperature diminishes by about 10 K from equator to pole, although recent IR data suggest a warm polar vortex exists in the upper atmosphere precisely at Saturn's poles. It is interesting that polar vortices occurring on Earth, Venus, Mars, and Jupiter are all colder than their surroundings, so these IR "hot spots" at the poles of Saturn appear rather unique and demand further study. In addition, the temperature structure of Saturn's troposphere is generally symmetrical with respect to equatorial regions. This is very curious, indeed, because with an obliquity of 26.7°, the expectation would be a marked seasonal effect on Saturn. But apparently the planet's response to variations in insolation is markedly slow at a distance of 9.54 AU from the Sun. It is worthwhile to mention also that, when the ring shadow obstructs the influx of sunlight in the equatorial regions of Saturn's globe, there appears to be no measurable effect on the planet's weather pattern!

In Figure 1.5, because Saturn does not have a definitive solid surface, the tropopause is the designated frame of reference at an altitude index of 0.0 km, and the top of the clouds reside about 50 km beneath the tropopause. A temperature inversion occurs at the level of the tropopause and in the still-higher stratosphere of Saturn heating of the upper atmosphere is due to the absorption of ultraviolet (UV) radiation from the sun by CH_4 (on Earth, an analogous absorption is accomplished by ozone). Above the stratosphere, the planet's ionosphere is exceedingly rarefied, containing mostly ionized hydrogen (H^+).

Layering occurs in the clouds of Saturn's atmosphere, as shown in Figure 1.5, and proceeding downward, there are layers of ammonia ice (NH_3) that condenses at temperatures lower than 145 K, ammonium hydrosulfide (NH_4HS) ice, and water (H_2O) ice. Unlike analogous regions on Jupiter, which amount to some 80 km in overall thickness, the three Saturnian layers (in addition to being individually thicker than their Jovian counterparts) are about 200 km in cumulative depth. The reason for this difference in cloud layer thickness is that more compression occurs in the atmosphere of Jupiter as a consequence of the planet's stronger gravitational attraction. The cloud layers are overlain, in the vicinity of the tropopause, by a blanket of haze formed by the interaction of sunlight with the upper Saturnian atmosphere. The deeper cloud layers of Saturn exhibit multiple hues resulting

Figure 1.6. Subtle, yet colorful, detail is visible in this excellent image taken with a 23.5-cm (9.25-in) Schmidt–Cassegrain (SCT) telescope and a Philips ToUcam webcam on December 11, 2004, at 02:26 Universal time (UT) in good seeing conditions by Damian Peach of Norfolk, UK. South is at the top of this image. (Credit: Damian Peach, ALPO Saturn Section.)

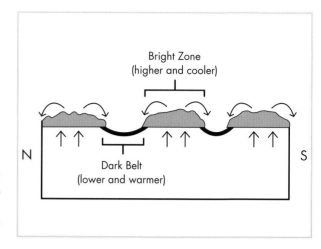

Bright Zone
(higher and cooler)

N

S

Dark Belt
(lower and warmer)

Figure 1.7. Cross-sectional view of convective structure of Saturn's belts and zones. (Credit: Julius L. Benton, Jr., ALPO Saturn Section.)

from essentially the same chemistry as on Jupiter, mostly due to reactions involving sulfur (S) and phosphorus (P), but organic compounds could possibly be involved, too. Unfortunately, on Saturn the global haze enveloping these regions obscures their colorful attributes, making the globe appear quite uniform much of the time. Anyone who has regularly viewed Saturn through a telescope will recall this distinctly less colorful aspect of the planet, but as mentioned earlier in this chapter, white or pale-yellowish zones and yellowish-to-tan belts (Fig. 1.6) are recognizable by persistent, careful observers, despite the haze layer. In addition, as a rule, the complex color changes that dramatically accompany storms on Jupiter do not commonly occur on Saturn.

Figure 1.7 is a simplified diagram of the convective structure of Saturn's belts and zones. It is clear that zones are bright, upwelling clouds in the cold upper atmosphere of the planet, their ascent driven by convective energy from below, while belts are more colorful descending regions exposing deeper, warmer areas of the atmosphere. Jet streams, moving rapidly eastward or westward on Saturn, induce complexities in the morphology of belts and zones because of Coriolis forces, which cause turbulent, rotating eddies to develop and evolve with time (Fig. 1.8).

Of all the atmospheric disturbances associated with Saturn's belts and zones, the zonal white spots (although rare) have traditionally been the most conspicuous, long-lasting features. The spectacular white appearance of these spots (as shown in Fig. 1.9) is probably due to NH_3 ice crystals, freshly formed as a plume of warm upwelling gas intrudes into the much colder upper cloud layers and not yet altered by chemical reactions that would induce coloration. With time, these bright spots can spread out and dissipate longitudinally within the zone(s) in which they appear.

In an analysis of Saturn's atmospheric wind patterns, it has become clear that the planet exhibits a predominantly equatorial eastward zonal flow, reaching velocities as much as 420 m/s at less than 30° latitude, almost four times the speed of the same region on Jupiter! Ascertaining exactly why there is such a difference between flow patterns on Saturn and Jupiter is a subject of intense current research. The velocity of the eastward flow diminishes somewhat as the distance north or south of the equator increases (e.g., 150 m/s at Saturnigraphic latitudes above 30°). The

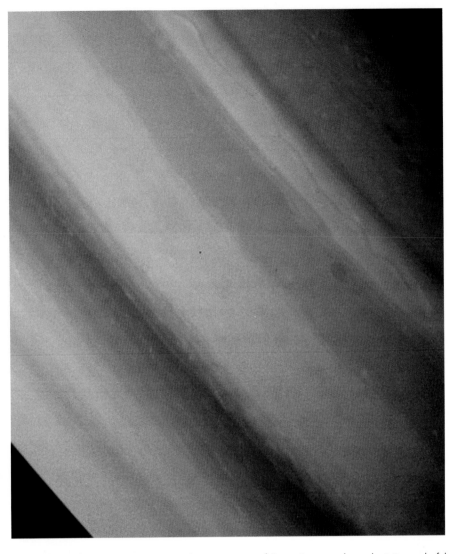

Figure 1.8. This is a representative close-up image of Saturn's atmosphere depicting colorful complexities in the morphology of belts and zones (taken by Voyager during the 1980s). North is at the top. (Credit: NASA, Jet Propulsion Laboratory, Pasadena, California.)

flow does not shift to a westward direction until latitude 40°N or S; then, in adjacent bands, an alternating eastward and westward circulation occurs toward the poles from these middle latitudes. Massive atmospheric flows and smaller scale disturbances or storms get most of their energy from a combination of convective motion from Saturn's interior and the planet's rapid rotation rate (Fig. 1.10). Unlike on Jupiter, there seems to be little or no correlation between the appearance of the visible cloud bands (bright zones and dark belts) on Saturn and the zonal easterly or westerly high-velocity winds at any given latitude or atmospheric level. Ultraviolet (UV) aurora emissions also occur near the top of Saturn's atmosphere within 12° or so of the poles (Fig. 1.11).

Figure 1.9. A brilliant white spot is clearly visible in the equatorial zone (EZ) of Saturn in this HST image taken on October 1998. Such features are due to NH_3–ice crystals forming and then rising upward into more frigid atmospheric regions and later spread longitudinally along the EZ. North appears at the top of the image. (Credit: NASA and Hubble space telescope [HST] Heritage Team [STScI/AURA].)

Figure 1.10. This stunning Cassini image shows that Saturn's atmosphere is an active and dynamic place, full of storms of varying dimensions and powerful winds. The smaller ones lie at the extreme threshold of vision of Earth-based observers. This view is of the planet's southern hemisphere and shows dark storms ringed by bright clouds. North is at the top. (Credit: NASA/Jet Propulsion Laboratory/Space Science Institute.)

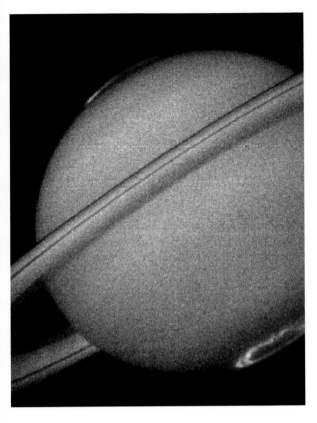

Figure 1.11. Saturn's spectacular UV aurora encircles the north and south poles, shown here as imaged by the space telescope imaging spectrograph (STIS) during October 1997 on board the Hubble space telescope (HST). These auroras occur when energetic wind from the sun sweeps over the planet, much like the Earth's nighttime auroral displays, but Saturn's auroras are only visible in UV light. North is up in this image. (Credit: NASA and Hubble space telescope [HST] Heritage Team [STScI/AURA].)

Saturn's Interior and Magnetosphere

We have already seen that the temperature at Saturn's cloud-tops is roughly 95 K, which is much warmer than the planet ought to be (i.e., 82 K is the expected temperature) if the cause is merely re-radiation of sunlight. Saturn has its own internal heat source, just like Jupiter, and it radiates considerably more energy than it absorbs from the sun. Massive Jupiter probably retained some of its original heat following gravitational contraction, but because of Saturn's comparatively lower mass and smaller size, its primordial heat would have long ago been radiated away. Therefore, the source of the planet's heat must be due to some other mechanism. Low-frequency emissions at radio wavelengths that emanate from Saturn fluctuate in a cyclic pattern inviting further investigations by spacecraft, and some of these outbursts may accompany massive lightning discharges in the atmosphere of the planet.

In a process that began in Saturn's remote past ($\sim 2.0 \times 10^9$ years ago), condensation of He into droplets takes place (and continues today) in the outer frigid layers of the planet's atmosphere, creating what is often called He precipitation. These droplets shower down as He rain through liquid H in the planet's interior to much lower levels, depleting the outer atmosphere of He. Gravitational forces act to compress the He, causing friction, and internal heat is released. From the beginning of this process $\sim 2.0 \times 10^9$ years ago, about 50% of the original He would have

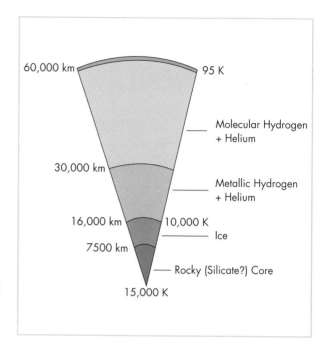

60,000 km

95 K

Molecular Hydrogen + Helium

30,000 km

Metallic Hydrogen + Helium

16,000 km

10,000 K

Ice

7500 km

Rocky (Silicate?) Core

15,000 K

Figure 1.12. A theoretical model of Saturn's interior. (Credit: Julius L. Benton, Jr., ALPO Saturn Section.)

fallen inward (Saturn has about 50% as much He as the atmospheres of Jupiter and the Sun). Figure 1.12 shows the internal structure of Saturn as derived from theoretical computer modeling.

Intuition suggests that Jupiter and Saturn ought to have similar interior compositions, and conjectural models indicate that Saturn's interior is composed of 74% H, 24% He, and 2% heavier elements, roughly approximating the composition of the sun. As can be seen in Figure 1.12, at a level of ~30,000 km beneath the clouds, a transition occurs from molecular H to metallic H, where the pressure is 3.0 Mbar (3.0×10^6 atmospheres). The point at which molecular H becomes metallic H lies much deeper inside Saturn than it does in Jupiter because of Saturn's comparatively lower mass and density (i.e., there is a slower increase in internal pressure inside Saturn as one progresses toward the center of the planet). At the center of the planet there may exist a core of silicate rock with a mass some 20 times that of the Earth.

In an earlier discussion, it was noted that Saturn has an internal rotation rate of $10^h39^m22^s$, and coupled with the planet's electrically conducting interior, a strong magnetic field results along with an extensive magnetosphere. Unlike Jupiter and the Earth, the planet Saturn has a magnetic field that is not inclined to the axis of rotation, with a strength that is ~1.0×10^3 that of the Earth's, but only 1/20th Jupiter's field intensity. Saturn's magnetic field, however, is still sufficient to produce a substantial magnetosphere and Earth-like radiation belts. The magnetosphere of Saturn, shown in Figure 1.13, extends 1.25×10^6 km in the direction of the sun, and it is large enough to encompass the ring system and many of the smaller innermost satellites. It also captures fewer particles than does the magnetosphere of Jupiter, probably due to the lack of an Io-like source of charged particles near Saturn, and the extensive ring system is very efficient in sweeping the inner magnetosphere clean of charged particles. At the outer edge of the rings, the

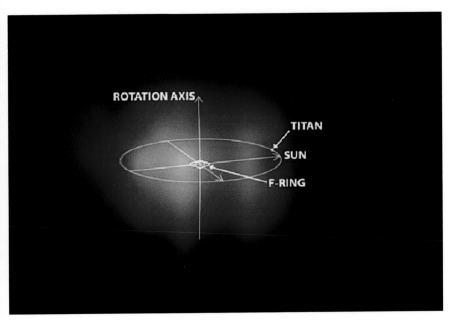

Figure 1.13. Saturn's magnetosphere, an envelope of charged particles surrounding some planets, including Earth, appears in this image taken by the Cassini spacecraft on June 21, 2004. Invisible to the human eye, the emission from these H atoms comes primarily from regions far from Saturn, well outside the ring system, and perhaps beyond the orbit of the largest moon Titan. (Credit: NASA/Jet Propulsion Laboratory/John Hopkins University.)

charged particle density increases rather dramatically, achieving a maximum at 3.0×10^5 km to 6.0×10^5 km from Saturn's center. In this region, the charged particles are tightly coupled to the rapidly rotating magnetic field, which gives rise to a plasma layer that is an estimated 1.2×10^5 km in thickness and extending as much as 9.0×10^5 km. At this point, the solar wind controls the extent of the Saturnian magnetosphere, where it may fluctuate between distances of 1.2×10^6 km and 1.8×10^5 km as the solar wind pressure changes. Note that the satellite Titan orbits Saturn in this region, and sometimes it may lie inside or outside the planet's magnetosphere depending on the intensity of the solar wind.

Saturn's Ring System

Natural scientific curiosity leads one to ponder why a ring of particles should exist around Saturn, how they got there in the first place, and what keeps them there. Two theories exist for the origin of Saturn's spectacular ring system, and a definitive answer to which theory is correct remains to be found. First, the rings may represent residual material from the formation of Saturn 4.6×10^9 years ago, or perhaps in an alternative view, a body roughly 250 km across might have drifted within the Roche limit of Saturn and was disrupted. It is interesting that Saturn's ring system has a total mass of 1.0×10^{16} kg, which is approximately the mass of a satellite of this size.

There is growing evidence that collisions among constituent ring particles would bring about the destruction of the ring system in a period considerably less than the age of the solar system. So, bodies making up the rings are obviously being replenished by some mechanism. Meteoroid bombardment of Saturn's satellites may be one possible source of this ring material, or maybe some relatively recent catastrophic event among the satellites contributed particles to the rings.

We have already established in a previous discussion that all of Saturn's rings reside within the planet's Roche limit, and we have seen that the orbital velocity of the bodies making up the rings decreases with increasing distance from the globe (e.g., a body on the inner edge of ring B orbits the planet in ~8^h, while one at the outer edge of ring A revolves around Saturn in ~14^h). Mutual collisions among ring particles have a tendency to sustain orbits that are co-planar and circular with prograde orbital motion, and the gravitational field of Saturn keeps the plane of the rings coincident with the planet's equatorial plane. As a result, the rings are extremely thin, no more than about 100 m thick, and brighter stars can be seen through most of the ring components from Earth in favorable conditions and with adequate telescopic aperture.

The Roche limit chiefly affects bodies that have enough mass to be held together by their own gravitational forces, not those where the dominant cohesive force is interatomic attraction. It is obvious, therefore, that very small bodies can exist within the Roche limit, and some of the largest ones are classified as moonlets. The vast majority of particles comprising the rings have a Bond albedo of 0.8, their high reflectivity suggesting a substantial icy content, and indeed, studies of the rings at IR wavelengths confirm that a major constituent is H_2O-ice, with an admixture of rocky material. With a surface temperature of 70 K, the ice making up ring particles is stable and does not evaporate (note that ring particles are somewhat shielded from solar radiation by other nearby particles, as well as sometimes by the shadow of Saturn's globe). The dominant size of ring particles is several centimeters across, with a range that encompasses objects with submillimeter proportions up to bodies tens of meters across.

Saturn's rings also exhibit an exquisite and immensely complex system of many thousands of ringlets with alternating regions of high and low density across their breadth, with a few gaps in addition to Cassini's and Encke's divisions (Fig. 1.14). Also, the extremely fine structure of the rings is not constant with time because gravitational interactions among ring particles cause spiral waves of varying density to emerge and evolve throughout the rings.

Recall that tiny moonlets, with diameters on the order of 10 to 25 km, can successfully reside without disruption inside Saturn's ring components (inside the Roche limit), and unlike the ringlets just described, the 20 or so true gaps in the rings probably arise from the clearing actions of these small bodies. For example, Saturn's 18th satellite, Pan, is located in Encke's division in ring A. The existence of small moonlets within the rings continues to remain as the most favored explanation for the presence of thin gaps. Small resonances among ring particles and these diminutive bodies fuel spiral density waves and sometimes distort ring components to a small degree.

Even Cassini's division is not completely devoid of particles, but the particle density in this gap is substantially lower than the concentration of material in either ring A or ring B. While the spacing within Cassini's division may arise from embedded tiny moonlets, the division as a whole arises from a 2:1 resonance of

Figure 1.14. The brightest part of Saturn's rings, curving from the upper right to the lower left in this image by the Cassini spacecraft on June 21, 2004, is the ring B (north is at the top of the image). Many bands throughout the B ring have a pronounced sandy color. Other color variations across the rings can be seen. Saturn's rings are made primarily of H_2O-ice, and since pure H_2O-ice is white, different colors in the rings may be due to different amounts of contamination by other materials such as rock or carbonaceous substances. (Credit: NASA/Jet Propulsion Laboratory/Space Science Institute.)

minute particles in the gap with the satellite Mimas. Resonant perturbations significantly decrease the concentration of material in Cassini's division.

The combined effect of resonances and the clearing influences of embedded moonlets bring about other ring phenomena. For instance, the sharpness of the outer edge of ring A is sustained by a 3:2 resonance between bodies orbiting in the outer edge of the ring and Mimas, plus the action of the small satellite Atlas (also in resonance with Mimas) helps ensure that material does not gradually escape outward. Furthermore, exclusive of Mimas, a 7:6 resonance of ring particles in the outer regions of ring A with the pair of satellites Janus and Epimetheus helps contribute to the sharply defined edge of ring A. It has been shown theoretically that, in the absence of such "favorable" perturbations, collisions and other interactions between ring particles would induce a gradual spreading out of the ring system.

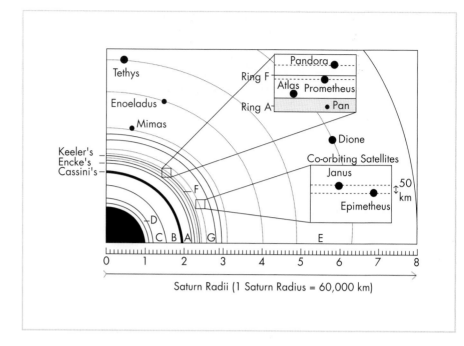

Figure 1.15. A detailed view of Saturn's rings. The view is high above Saturn's north pole, looking down on the rings. (Credit: Julius L. Benton, Jr., ALPO Saturn Section.)

Look at the diagram in Figure 1.15, which presents a detailed view of Saturn's rings, showing components, embedded satellites, known gaps, etc., while Table 1.3 lists basic data for Saturn's ring components and divisions.

The following subsections describe each ring component and any associated gaps or divisions in a little more detail in order of distance from the center of Saturn.

Table 1.3. Saturn's ring system: basic data			
Name	Inner radius (km)*	Outer radius (km)*	Width (km)*
Ring D	67,000	74,510	7,510
Guerin gap	74,510	74,658	148
Ring C	74,658	92,000	17,342
Maxwell division	87,500	88,000	500
Ring B	92,000	117,500	25,500
Cassini's division	117,680	120,600	4,800
Huygens gap	117,680	122,200	~4,520
Ring A	122,200	136,800	14,600
Encke's complex	126,430	129,940	3,500
Keeler's division	133,580	133,905	~325
Ring F	140,210	140,600	~390
Ring G	165,800	173,800	8,000
Ring E	180,000	480,000	300,000

*Distances are in kilometers from Saturn's center.

Figure 1.16. Voyager 2 took this picture of Saturn's exceedingly faint inner D ring August 25, 1999. Ring D is very tenuous and has an extremely small optical depth. North is up in the image. (Credit: NASA, Jet Propulsion Laboratory, Pasadena, California.)

Ring D

Ring D appears to exhibit no well-defined inner edge. It is actually a very faint series of ringlets, at least one of which is eccentric (Fig. 1.16). It may begin just above Saturn's cloud-tops at a distance of 67,000 km from the center of the planet (~7000 km above the clouds). Ring D is extremely dark, with relatively few particles in it, and because of the overwhelming glare of Saturn's globe and the rest of the rings, it is not considered to be visible from Earth (except maybe indirectly when stars are occulted by the rings). Ring D has an orbital period (center of ring component) of roughly 5.5^h. At the outer edge of ring D (at 74,510 km) is the exceedingly narrow Guerin gap, about 148 km wide.

Ring C

Sometimes called the Crape ring (typically when it is visible in front of the globe of Saturn), ring C begins at the outer terminus of ring D and the Guerin gap, some

Figure 1.17. This view of Saturn's outer C ring by the Cassini spacecraft on October 29, 2004, shows the extreme variations in brightness, along with subtle, large-scale wavy variations (north is up in this view). The notably dark Maxwell gap (near upper right) contains the bright, narrow, and eccentric Maxwell ringlet. (Credit: NASA/Jet Propulsion Laboratory/Space Science Institute.)

74,658 km outward from Saturn's center (some 24,500 km above the cloud-tops), extending to the inner edge of ring B (at 92,000 km from the planet's center). Ring C has an orbital period of 5.8^h (inner edge) to 7.9^h (outer edge). In Earth-based telescopes, ring C is quite dark and more readily detected at the ansae when the rings are wide open.

Across its width of some 17,342 km, ring C has numerous concentric ringlets not appreciably altered in position by resonance factors, as clearly shown in Figure 1.17. The Maxwell division, which is 500 km wide, extends from 87,500 km out to 88,000 km from the center of Saturn in ring C. Extremely narrow ringlets apparently occur within this gap, a few of which appear to be eccentric. The eccentricity of such ringlets is probably the result of dynamic forces imposed on them by small moonlets, or caused by gravitational interactions producing spiral density waves.

Ring B and Cassini's Division

Brightest of all of the ring components, ring B begins at the outer edge of ring C at 92,000 km (where no apparent gap exists) and extends for some 25,500 km out to a distance of 117,500 km from the center of Saturn. The inner edge of ring B has an orbital period of 7.9[h], while the outer edge of the component orbits in 12[h]. Visual observers are well aware that the inner two thirds of ring B is usually dimmer than the outer third. Ring B is composed of thousands of ringlets and numerous minor divisions, with a higher number of gaps in the inner two thirds of the ring (Fig. 1.18). While ring B seems to be the most opaque component from Earth, it seems thickest in the outer third and at its innermost edge, and it has been noted that the

Figure 1.18. In this close-up image of ring B and Cassini's division made by the Cassini spacecraft on October 29, 2004, many subtle wavelike patterns are visible, along with hundreds of narrow features resembling "grooves" of a phonograph record. There is a noticeable abrupt change in overall brightness beyond the dark gap near the right. To the left of the gap is the outer ring B, with its sharp edge maintained by a strong gravitational resonance with the satellite Mimas. To the right of Huygen's gap are the plateau-like bands of Cassini's Division. The narrow ringlet within the gap is called the Huygens ringlet. North is at the top. (Credit: NASA/Jet Propulsion Laboratory/Space Science Institute.)

thickness of the overall ring system reaches a peak of ~100 km in ring B. Extraplanar particles of rarefied H occur above and below rings A and B, but it is questionable whether this extremely tenuous haze can be seen from Earth at edgewise ring presentations. The fragments making up ring B range from several centimeters up to a few meters across, and spectroscopy reveals that they probably have a slightly redder appearance (different composition?) than particles in adjacent rings C and A.

Dusky radial spokes, which may be up to 20,000 km in linear extent, appear from time to time within ring B, as seen in Figure 1.19. These transient radial features are composed of micrometer-size dust that levitates several tens of meters above and below the ring plane due to electrostatic forces, generated most likely by particle collisions within the rings. As they co-rotate with the rings through one revolution, dispersal of the charged fields gradually takes place, resulting in disappearance of the radial spokes. Although they are not permanent features, the radial

Figure 1.19. Evident here in this Voyager 2 image (north is at the top right) taken on August. 3, 1981, are numerous "spoke" features in Saturn's ring B. Their very sharp, narrow appearance suggests short formation times, and electromagnetic forces are probably in some way responsible for these filamentary features. (Credit: NASA, Jet Propulsion Laboratory, Pasadena, California.)

spokes associated with ring B are probably routinely recurring phenomena. It is also thought that these features very rarely occur in ring A and probably not at all in dusky ring C.

At a distance of 117,680 km from Saturn's center is the inner edge of the well-known Cassini's division, strikingly depicted in Figure 1.18. It spans about 4800 km out to 120,600 km, but out at 117,680 km is a subdivision in the material known as Huygen's gap, which roughly terminates near the inner edge of ring A at 122,200 km from the center of Saturn. In Earth-based telescopes in good viewing conditions, Cassini's looks almost black, but observations of stars as they pass behind Cassini's division show tiny fluctuations in brightness. This occurs because extremely small debris, similar to particles making up ring C, exists inside this gap, but most of it has been cleared by the 2:1 resonance with Saturn's nearby satellite Mimas.

Figure 1.20. In this view of Saturn's ring system taken by the Cassini spacecraft on August 27, 2004, the diaphanous C ring appears at the upper right, followed by the multihued B ring (north is up in this view). Next, Cassini's division separates rings A and B. The outer edge of ring B, which forms the inner boundary of Cassini's division, is maintained by gravitational resonance with the satellite Mimas. Near the outer edge of ring A is Encke's gap, while Keeler's gap is barely visible. The faint, thread-like ring F is discernible just beyond the main rings. (Credit: NASA/Jet Propulsion Laboratory/Space Science Institute.)

Ring A, Encke's Division, and Keeler's Gap

Considerably duller than ring B, ring A begins approximately at the outer boundary of Cassini's division at 122,200 km from the center of Saturn, extending outward for 14,600 km, and then ending at 136,800 km. The inner and outer edges of ring A have orbital periods of 12^h and 14.4^h, respectively. Although Saturn's rings are generally uniform longitudinally, ring A sometimes displays an azimuthal brightness asymmetry near the ansae (the east and west regions of the rings where they are farthest from Saturn's globe), attributed most likely to transient "density wakes" caused by clumping of particles due to gravitational forces.

As Figure 1.20 shows, ring A exhibits minor divisions and ringlets just like ring B, and the more distinct Encke's complex some 3500 km wide occurs from 126,430 km out to 129,940 km from Saturn's center (from 29% to 53% of the distance

Figure 1.21. An intriguing knotted ringlet within Encke's division (located within ring A) is of interest in this image made by the Cassini spacecraft on October 29, 2004. The tiny moon Pan orbits within Encke's division and maintains it. Many waves produced by orbiting moons are also visible. View is from the north of the ring plane. (Credit: NASA/Jet Propulsion Laboratory/Space Science Institute.)

between the inner and outer boundaries of ring A). Encke's complex can be seen in moderate-size telescopes from Earth in favorable conditions, but it is never as conspicuous as Cassini's, and it is made up of numerous ringlets, a few of which appear to be eccentric (Fig. 1.21). About 80% of the distance out in ring A, at a distance of 133,580 km from Saturn's center, is the 325-km-wide Keeler gap (marginally visible in Fig. 1.20), which also has several faint ringlets within it. We have already discussed earlier how the satellites Mimas, Atlas (in resonance with Mimas), Janus, and Epimetheus conspire in a complex way to keep the outer edge of ring A very sharp.

Ring F

Perhaps the most bizarre Saturnian ring component is ring F (Fig. 1.22), which has a distance range of 140,210 km to 140,600 km from the center of Saturn, with a

Figure 1.22. Saturn's faint, narrow, "braided" ring F appears in this Cassini image during the spacecraft's initial approach to Saturn in mid-2004. North is up in this image. (Credit: NASA/Jet Propulsion Laboratory/Space Science Institute.)

width that spans 30 to 500 km and an orbital period of nearly 15^h. It is narrow and faint, lying just barely inside Saturn's Roche limit, roughly 3400 km outside ring A. Ring F is very slightly eccentric, consisting of a number of strands intertwined. The thinness and "braided" longitudinal structure of the ring seems linked to shepherd satellites that orbit on either side of ring F, but the mechanism for production of these irregular variations is still unclear. Shepherding satellites may also have something to do with the existence of eccentric ringlets seen in gaps in other major ring components. Refer back to Figure 1.15 and note that two tiny, dark moons named Pandora and Prometheus orbit roughly 1000 km on either side of ring F, and despite their small dimensions (each no greater than about 150 km across), their gravitational influence on particles in ring F maintains it as a tightly compressed thin component, as shown in Figure 1.23.

Figure 1.23. Two of Saturn's moons, Prometheus (situated inside ring F) and Pandora (just outside ring F), are seen here shepherding the planet's narrow ring F in this Cassini image made on May 1, 2004 (north is up in this image). Prometheus overtakes Pandora in orbit around Saturn about every 25^d. Slightly above the pair and to the right is another satellite, Epimetheus. Credit: NASA/Jet Propulsion Laboratory/Space Science Institute.)

Ring G

In order of increasing distance from Saturn's center, the next component is the extremely faint ring G extending from 165,800 km out to 173,800 km (width of 8000 km). Ring G is situated between the orbits of Mimas and co-orbiting satellites Janus and Epimetheus (Fig. 1.15) with an orbital period of about 20^h. This feeble, optically thin ring component shows no internal structure.

Ring E

At an inner radius of 180,000 km from the center of Saturn out to about 480,000 km (width of 300,000 km) is ring E, the last and outermost of the currently known ring components. This ethereal ring component is so extensive that it can sometimes be seen by Earth-based observers near edgewise ring presentations. The orbital period of ring E ranges from 22^h at its inner edge to 95^h (3.96^d) at its extreme outer edge. The inner portion of the ring is a little brighter than the outer part and this brighter region of ring E lies just barely within the orbit of the satellite Enceladus. It has been suggested that meteoroid erosion and volcanism occurring on Enceladus may be a source of some of the material found in ring E.

The Satellites of Saturn

Although an extensive family of at least 33 known satellites (approximately 17 more small moonlets have been identified recently by spacecraft near Saturn) accompanies Saturn, giant Titan overwhelmingly dominates the scene and is the only moon in the solar system with a dense atmosphere. Probably all of Saturn's moons contain a relatively high percentage of H_2O-ice (as well as an admixture of carbonaceous dirt on some), but there are a few interesting morphological differences among these exotic worlds. For example, Mimas has a huge crater occupying nearly a third of the satellite's diameter; the surface of Enceladus has a combination of smooth, fractured, and cratered terrain; and Rhea has an old, heavily cratered surface. Tethys and Dione have nearly the same diameters, but Dione is a little denser. Hyperion is biscuit-shaped, and irregular dimensions characterize at least nine of Saturn's other smaller moons. The leading hemisphere of Iapetus is dark, while the trailing hemisphere is nearly five times as bright, and Phoebe orbits backward (in a retrograde fashion) around Saturn! Some of the smallest satellites act as shepherd moons and some co-orbit with larger satellites. Table 1.4 lists the satellites of Saturn, along with important data about each one of them, Figure 1.24 is a very striking colorful montage of some of Saturn's satellites as imaged by passing spacecraft.

Seven bodies among the nine classical satellites of Saturn listed in Table 1.4 have roughly circular orbits that are inclined by ~1.5° to the planet's equatorial plane (and that coincide also with the plane of the ring system). The two notable exceptions are the orbit of Iapetus, which is inclined to the equatorial plane of Saturn by 14.7°, and the path of Phoebe, which has an orbital inclination of 174.8° and retrograde revolution. The following subsections discuss some of the more significant characteristics of each of the Saturnian satellites.

Table 1.4. Basic data on Saturn's satellites

Number	Name	Size (km)	a (km)	i (°)	e	Orbital period (days)	Albedo	Magnitude (V_o or R)	Surface materials
Regular satellites									
S18	Pan	20	133,600	0.00	0.00	0.58	0.5	19.4	H_2O-ice (?)
S15	Atlas	32	137,700	0.00	0.00	0.60	0.4	19.0	Dirty H_2O-ice (?)
S16	Prometheus	100	139,400	0.00	0.00	0.61	0.6	15.8	H_2O-ice (?)
S17	Pandora	84	141,700	0.00	0.00	0.63	0.5	16.4	H_2O-ice (?)
S11	Epimetheus	119	151,400	0.34	0.02	0.69	0.5	15.6	Dirty H_2O-ice (?)
S10	Janus	178	151,500	0.17	0.01	0.70	0.6	14.4	Dirty H_2O-ice (?)
S1	Mimas*	397	185,600	1.57	0.02	0.94	0.6	12.8	H_2O-ice
S32	S/2004 S1	~6	194,000	—	—	1.01	0.06	—	H_2O-ice (?)
S33	S/2004 S2	~8	211,000	—	—	1.14	0.06	—	H_2O-ice (?)
S2	Enceladus*	499	238,100	0.01	0.00	1.37	1.0	11.8	H_2O-ice
S13	Telesto	24	294,700	1.16	0.00	1.89	1.0	18.5	Dirty H_2O-ice
S3	Tethys*	1,060	294,700	0.17	0.00	1.89	0.8	10.2	H_2O-ice
S14	Calypso	19	294,700	1.47	0.00	1.89	0.7	18.7	H_2O-ice
S4	Dione*	1,118	377,400	0.00	0.00	2.74	0.6	10.4	Dirty H_2O-ice
S12	Helene	32	377,400	0.21	0.00	2.74	0.6	18.4	Dirty H_2O-ice
S5	Rhea*	1,528	527,100	0.33	0.00	4.52	0.6	9.6	Dirty H_2O-ice
S6	Titan*	5,150	1.22×10^6	1.63	0.03	15.95	0.2	8.4	CH_4; N_2 atmosphere
S7	Hyperion*	350 × 200	1.46×10^6	0.57	0.02	21.28	0.3	14.4	Dirty H_2O-ice
S8	Iapetus*	1,436	3.56×10^6	7.57	0.03	79.33	0.6	11.0	Dark soil; dirty H_2O-ice
Irregular satellites — prograde group									
S24	Kiviuq	~14	1.14×10^7	46.16	0.33	449.2	0.06	22.0R	Dirty H_2O ice?
S22	Ijiraq	~10	1.15×10^7	46.74	0.32	451.5	0.06	22.6R	Dirty H_2O ice?
S20	Paaliaq	~19	1.52×10^7	45.13	0.36	686.9	0.06	21.3R	Dirty H_2O ice?
S26	Albiorix	~26	1.64×10^7	33.98	0.48	783.5	0.06	20.5R	Dirty H_2O ice?

Table 1.4. Continued

Number	Name	Size (km)	a (km)	i (°)	e	Orbital period (days)	Albedo	Magnitude (V₀ or R)	Surface materials
Irregular satellites—prograde group									
S28	Erriapo	~9	1.76×10^7	34.45	0.47	871.9	0.06	23.0R	Dirty H_2O ice?
S29	Siarnaq	~32	1.82×10^7	45.56	0.30	893.1	0.06	20.1R	Dirty H_2O ice?
S21	Tarvos	~13	1.83×10^7	33.51	0.54	925.6	0.06	22.1R	Dirty H_2O ice?
Irregular satellites—retrograde group									
S9	Phoebe*	220	1.29×10^7	174.8	0.16	548.2	0.08	16.4	Dirty H_2O ice
S27	Skadi	~6	1.56×10^7	152.7	0.27	728.9	0.06	23.6R	Dirty H_2O ice?
S25	Mundilfari	~6	1.87×10^7	167.5	0.21	951.4	0.06	23.8R	Dirty H_2O ice?
S31	S/2003 S1	~7	1.87×10^7	134.6	0.35	956.2	0.06	24.0R	Dirty H_2O ice?
S23	Suttung	~6	1.95×10^7	175.8	0.11	1016.3	0.06	23.9R	Dirty H_2O ice?
S30	Thrym	~6	2.04×10^7	175.8	0.47	1086.9	0.06	23.9R	Dirty H_2O ice?
S19	Ymir	~16	2.31×10^7	173.1	0.33	1312.4	0.06	21.7R	Dirty H_2O ice?

* Classical satellites of Saturn.
Size: diameter of the satellite in kilometers (km).
a, mean semi-major axis in kilometers (km).
i, mean inclination in degrees (°), where inclinations *bg90° indicate retrograde orbits.
e, mean eccentricity.
Orbital period: time for one revolution around Saturn in days.
Albedo: visual geometric albedo (mean reflectivity).
Magnitude: V₀, mean opposition magnitude; R, red magnitude.

The Satellites of Saturn

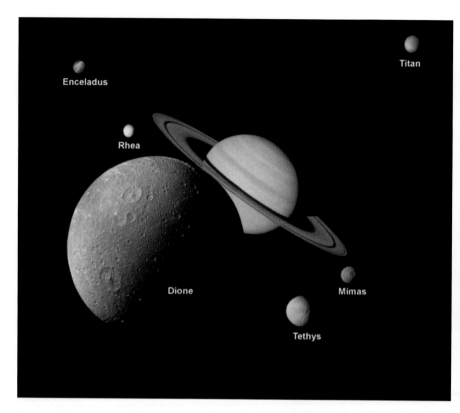

Figure 1.24. This montage of the Saturnian system was prepared from an assemblage of images taken by the Voyager 1 spacecraft during its flyby during November 1980. This artist's view shows Dione in the forefront, Saturn rising behind, Tethys and Mimas fading in the distance to the right, Enceladus and Rhea off Saturn's rings to the left, and Titan in its distant orbit at the top. North appears at the top of the image. (Credit: NASA, Jet Propulsion Laboratory, Pasadena, California.)

Titan

Titan is a truly impressive object, with the distinction of being the only satellite in our solar system enshrouded by a thick atmosphere. It has a diameter of 5150 km, placing it ahead of Mercury (4878 km), and in the hierarchy of planetary satellites, it ranks just behind Jupiter's Ganymede (5268 km). Titan displays a distinct reddish-orange hue (Fig. 1.25), attributed to reactions of methane (CH_4) with solar UV radiation that yield complex hydrocarbons and polymers. The result is photochemical "smog," but the CH_4-aerosol combination accounts for only ~10% of the atmosphere.

Indeed, remote sensing by spacecraft reveals the well-known global detached haze layer caused by photochemical reactions hundreds of kilometers above Titan's surface, visible as a thin ring of bright material enveloping the entire satellite, as depicted in Figure 1.26.

The main constituent of Titan's atmosphere is 90% to 97% nitrogen (N_2), with traces of H_2, acetylene (C_2H_2), ethane (C_2H_6), ethylene (C_2H_4), hydrogen cyanide (HCN), and carbon monoxide (CO). Under the influence of sunlight, elaborate

Figure 1.25. This natural color image was taken by the Cassini spacecraft as it sped silently past Titan on July 2, 2004. Titan's south polar Rregion appears in this half-phase view (north is up in this image). (Credit: NASA/Jet Propulsion Laboratory/Space Science Institute.)

chemical reactions probably occur at higher atmospheric levels, producing a "petrochemical rain." According to some theorists, its surface may have some regions covered with a substance akin to asphalt! Near ground level the air pressure is higher than the Earth's by a factor of 50% or more, and the surface temperature of Titan is a very frigid 93 K, varying only about ±3 K from the satellite's equator to its poles. Theoretical models predict CH_4 or N_2 rain, but liquid CH_4 may evaporate before reaching the surface. N_2 rain, if it exists, would be more likely to form puddles on the surface, but lakes of liquid CH_4 or even ethane (C_2H_6) are possible. The extremely cold environment of Titan, in fact, may cause CH_4 or N_2 to play a similar role there as H_2O does in its different phases on Earth. Near-IR studies by spacecraft using filters sensitive to CH_4 reveal streaks on Titan's surface that may be the result of movement of a fluid over the surface, such as wind, hydrocarbon liquids, or a migrating ice sheet like a glacier (Fig.1.27).

Because of the optically impenetrable dense CH_4 haze enveloping Titan, the actual surface is hidden from view, but IR studies and remote sensing by passing spacecraft at 9380A[o] hint that the satellite may have quite a few relief features,

Figure 1.26. A global detached haze layer and discrete cloud-like features high above Titan's northern terminator (day–night transition line) are visible in this Cassini spacecraft image acquired on October 24, 2004, using a near-UV filter that is sensitive to scattering of small particles. This full-disk view is a natural colorized version of the original UV image. North is at the top of the image. (Credit: NASA/Jet Propulsion Laboratory/Space Science Institute.)

most notably the recently discovered "continent size" feature known as Xanadu, intermingled with depressed terrain (Fig. 1.28).

The most up-to-date spacecraft images of Titan's surface in early 2005 reveal a weird orange terrain that looks that way because of weak sunlight passing through the reddish-orange atmospheric smog enveloping the satellite. Pebble-size "rocks" up to 15 cm across, probably consisting of a blend of H_2O-ice and various frozen hydrocarbons, appear here and there, and the same images show hints of a ground fog, plus what looks like a shoreline bordering features reminiscent of wide, dark seas (Figs. 1.29 and 1.30). These "seas" are presumably moist bare ground with liquid CH_4 just beneath a surface that appears a bit darker than theorists anticipated. At the base of many of the rounded rocks are signs of erosion, in all probability due to some form of fluvial activity. Other discoveries include a gigantic

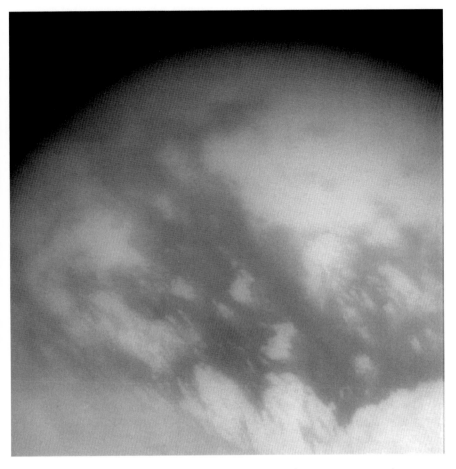

Figure 1.27. Streaks of surface material in Titan's equatorial region are seen in the east-west direction (upper left to lower right with north at the top) in this near-IR image taken by Cassini on October 24, 2004 (wavelengths sensitive to CH_4). They may be caused by fluid motion over the surface (i.e., hydrocarbon liquids or migrating ice sheets). The large-scale streaks are likely due to winds. (Credit: NASA/Jet Propulsion Laboratory/Space Science Institute.)

ringed basin some 440 km across, called Circus Maximus, and channels up to 200 km long that appear to emanate from the slopes of the crater. The channels are probably the result of flowing liquid CH_4, considering the extremely frigid environment of Titan's surface. A bright ejecta blanket surrounds some impact craters, and spacecraft measurements suggest wind speeds at Titan's surface near 7 cm per hour.

The bulk density of Titan is 1.880 kg/m^3, with an interior structure that in all probability consists of a silicate core overlain by a thick mantle of H_2O-ice. The absence of an intrinsic magnetic field suggests there is probably no liquid component in Titan's interior.

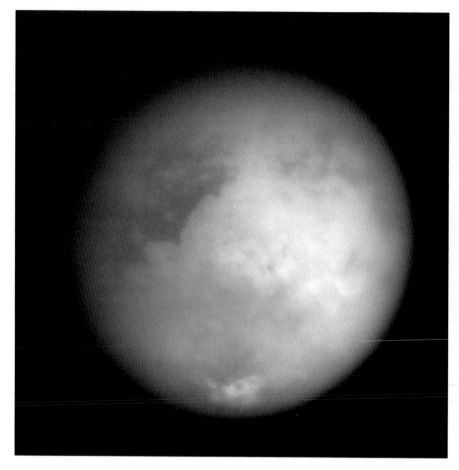

Figure 1.28. This image taken on October 24, 2004, reveals Titan's bright "continent-sized" feature dubbed Xanadu at 9380Å (a near-IR wavelength region at which the surface is easily detected). The origin and geography of Xanadu merits further study. Bright features near Titan's south pole (bottom) are unquestionably clouds. (Credit: NASA/Jet Propulsion Laboratory/Space Science Institute.)

The Midsize Moons: Mimas, Enceladus, Tethys, Dione, Rhea, and Iapetus

Mimas is 397 km in diameter with a bulk density of 1190 kg/m^3. It is the smallest and innermost of the medium-sized satellites, and it has synchronous rotation (i.e., always keeps the same face toward Saturn). It is heavily cratered, and there is no evidence that any kind of tectonic or volcanic resurfacing has taken place since Mimas formed. The bright surface of Mimas is mainly due to the presence of H_2O-ice, and its low density implies a similar interior composition. The most prominent surface feature, visible in Figure 1.31, is a gigantic impact crater named Herschel, which is 130 km across (about one-third the diameter of Mimas) with walls and a central peak rising about 5 to 6 km from its floor. The collision that formed this crater in all probability nearly disintegrated the satellite! The smaller craters on

Figure 1.29. The Cassini-Huygens probe took this image on January 14, 2005, as it approached landing on Titan. Pebble-size lumps of H_2O-ice mixed with frozen hydrocarbons are visible in this view, ranging from 4 to 15 cm across, with some evidence of erosion at their bases. Titan's surface is darker than originally expected, and the orange hue is a result of feeble sunlight passing through Titan's thick, smoggy atmosphere. (Credit: NASA/European Space Agency/Italian Space Agency/Jet Propulsion Laboratory.)

Mimas (those between approximately 2 and 20 km across) are bowl-shaped and are a bit deeper than craters of similar size on the Galilean satellites or the moon. Beyond crater diameters of 20 km, central peaks are common, but few craters exist larger than 50 km across. Resonance interactions of Mimas with ring particles, whereby it contributes to the formation of Cassini's division and the sharpness of the outer edge of ring A, is a consequence of its close relative proximity to the rings.

Orbiting just outside Mimas, Enceladus is 499 km in diameter with a bulk density 1200 kg/m^3 and is gravitationally locked into synchronous rotation just

Figure 1.30. This 360° mosaic of Titan's surface was produced from images taken on January 14, 2005, by the Cassini-Huygens probe as it landed on Titan. To the left, behind Huygens, notice the boundary between light and dark areas occupying a surface darker than what theorists initially expected. White striations seen near this boundary may be ground "fog" since these streaks were not visible from higher altitudes. Wind speeds of 6 to 7 km per hour were also detected. (Credit: NASA/European Space Agency/Italian Space Agency/Jet Propulsion Laboratory.)

Figure 1.31. The cratered surface Saturn's moon Mimas is seen in this image taken by Cassini on January 16, 2005. Abundant small craters caused by impacts of cosmic debris are everywhere, attesting to the ancient age of Mimas's surface; the largest crater, known as Herschel, is 130 km across and displays a prominent central peak. North is up in this photo. (Credit: NASA/Jet Propulsion Laboratory/Space Science Institute.)

Figure 1.32. Moderately cratered areas on Enceladus are transected by strips of younger grooved terrain in this Voyager 2 image made on August 25, 1981. Since younger terrain seems to have consumed portions of craters like those near the bottom center of the image, and crustal deformations have produced grooves and streaks, internal melting likely occurred in the past. The largest crater visible here is about 35 km across. North is toward the top of the image. (Credit: NASA, Jet Propulsion Laboratory, Pasadena, California.)

like its neighbor. Because it is relatively close to Mimas, one might assume that the two bodies are alike in history and morphology, but rather striking dissimilarities exist. The most notable difference is that Enceladus is sparsely cratered, which strongly suggests that preexisting impact craters have been obliterated by H_2O "lava" flows, although no actual volcanoes have been observed (Fig. 1.32). Indirect evidence of such upheavals might be the increased concentration of thin icy particles in ring E nearest Enceladus. Could volcanism on this satellite be replenishing material in ring E, a ring component that seems to be subject to depletion of material by the solar wind? Some investigators postulate that, because of the high reflectivity of Enceladus (it has the most reflective surface in the solar system), there might be an extensive crystalline layer of pure H_2O-ice on its surface. The few

Figure 1.33. Two Voyager 2 images of Tethys taken in August 1981. The larger image shows heavily cratered areas and regions with fewer craters, suggesting early internal activity partially resurfaced the older terrain. A huge trench system, Ithaca Chasma, girdles nearly 75% of Tethys's circumference, perhaps formed when a warmer H_2O-ice rich interior expanded as it froze. The smaller image to the lower right shows the opposite hemisphere of Tethys and the huge 400 km diameter crater Odysseus, flattened out over time because the moon's interior was warmer and softer. North is up in this view. (Credit: NASA, Jet Propulsion Laboratory, Pasadena, California.)

craters that do exist occur in specific areas and range in size from bowl-shaped scars ~10 km in diameter to several around 30 km across with rounded central peaks. Other features include faults and ridges that are indicative of crustal movements. The source of Enceladus' apparent internal activity remains a mystery, because there seems to be insufficient tidal stress to drive H_2O volcanism. Resonant interactions, however, may be enough to cause flexure of the small satellite, generating sufficient internal heat to melt subsurface H_2O-ice and give rise to liquid eruptions onto the surface, quickly becoming bright new snow.

Tethys is 1060 km in diameter and has a bulk density 1300 kg/m³. It has an abundance of similar-sized small impact craters like Mimas, and it sports a colossal crater, Odysseus, that is some 400 km across (roughly 40% the diameter of Tethys). Relative to the size of the body on which it exists, Odysseus is the largest crater in the solar system! On the opposite side of Tethys from Odysseus and extending over the north pole of Tethys, down across the equator, and nearly to its south pole, is Ithaca Chasma. It is a canyon or groove with a mean width of 100 km and a depth of about 5 km as it encircles about 75% of the satellite (see the two images of Tethys in Fig. 1.33) Whether or not the formation of Odysseus is connected to the origin of Ithaca Chasma remains unclear. The low density of Tethys implies a significant H_2O-ice content, and the presence of plains in some areas suggests localized resurfacing activity. Tethys, like Mimas and Enceladus, has synchronous rotation.

Dione is the next midsized world orbiting Saturn beyond Tethys in captured rotation, with a diameter of 1118 km and a bulk density 1400 kg/m^3. The brightness of the surface of Dione is quite irregular, with regions that are relatively bright, contrasted with others that are a bit darker. Although Dione is profusely cratered like much of Tethys, there are regions of smooth plains with brilliant, wispy terrain, which may be surface ruptures originating from internal stresses, allowing upheaval of bright, pure H_2O lava (Fig. 1.34). Craters are typically about 30 km in diameter on Dione, but the few largest ones (~170 km across) have rather obvious central peaks and floors that are not as deep as those on Tethys. With a slightly higher bulk density than Tethys as a clue, the interior of Dione may have a higher percentage of silicate material relative to H_2O-ice. Also, the incidence of radiogenic minerals inside Dione may be a source of internal heat.

Saturn's second-largest moon, Rhea is 1528 km in diameter and has a bulk density 1330 kg/m^3. It is another synchronous-rotating, heavily cratered body, with

Figure 1.34. Numerous large-impact craters, bright rays, and brilliant ridges and valleys are seen in this view of Dione taken by Voyager 1 on November 12, 1980. Irregular valleys that represent old fault troughs, degraded by impacts, are also present. North is at the top of the picture. (Credit: NASA, Jet Propulsion Laboratory, Pasadena, California.)

Figure 1.35. The heavily cratered ancient surface of Rhea appears in this Voyager 1 image taken on November 12, 1980. Craters 2.5 km across are abundant, but many areas of Rhea's surface are lacking in 100 km or larger craters, indicating a change occurred in the nature of impacting bodies over time and an early period of surface activity. North is up in this image. (Credit: NASA, Jet Propulsion Laboratory, Pasadena, California.)

a surface of H_2O-ice (Fig. 1.35). The crater density on Rhea is analogous to that of the lunar highlands, suggesting that there has been little or no volcanic activity to modify the surface of the satellite during its lifetime. Sizes of craters approach 70 km from rim to rim, with many of the larger ones displaying central peaks. A very puzzling aspect of Rhea is its bright, wispy terrain confined to its trailing hemisphere. As with seemingly matching light-colored streaks on Dione, Rhea may have experienced a period in its history that allowed the release of H_2O from beneath its surface, which formed a highly reflective frost layer. Models support the view that the interior of Rhea is silicate and H_2O-ice in almost equal proportions, with denser rocky material having migrated to the core during the satellite's differentiation phases.

The last of the middle-sized satellites is the strange world Iapetus, which exhibits two distinctly different hemispheres. It is 1436 km in diameter and has a bulk density of 1200 kg/m³. As Iapetus orbits Saturn in its eccentric and highly inclined orbit of 14.7°, it keeps the same face always turned toward the planet, and the satellite's leading hemisphere is considerably darker than the trailing one. Iapetus is densely cratered, but the brighter trailing face of Iapetus (albedo of 0.5) is covered by extensive amounts of H_2O-ice, while the leading dark hemisphere (albedo of 0.02) is H_2O-ice overlain by dark (carbonaceous?) soil, as depicted in Figure 1.36. Because of extensive surface cratering, the boundary between the light and dark regions of Iapetus is sharply defined and irregular. It may be that the dark leading-hemisphere deposits originated on Iapetus itself, but it is also possible that dark material "sand-blasted" off Phoebe by meteoroids eventually ended up on the leading side of the satellite. The most prominent feature on Iapetus's icy surface is Cassini Regio, a dark, heavily cratered area that occupies nearly its entire leading hemisphere. The origin of Cassini Regio remains uncertain, but many theorists feel that its dark material possibly erupted onto the icy surface of Iapetus from the interior, or the dark terrain may be accumulated debris ejected by impact events on the darker outer satellites of Saturn. In addition to Cassini Regio, a truly remarkable topographic ridge runs across Iapetus coinciding roughly with the satellite's geographic equator (this feature is quite conspicuous in Fig. 1.36). This long ridge may be a narrow strip of mountainous terrain that folded upward, or perhaps it represents an extended "crack" through which material erupted from the interior of Iapetus onto its surface. The interior of the satellite is probably mostly H_2O-ice.

Figure 1.36. Saturn's outermost large moon, Iapetus, has contrasting bright, heavily cratered icy terrain on the trailing hemisphere and dark terrain on the leading side. Cassini flew past Iapetus on December 31, 2004, capturing four images that were combined to show a dark, heavily cratered region, called Cassini Regio that covers nearly the entire leading hemisphere of Iapetus. Notice also the prominent ridge that runs almost exactly along the geographic equator of Iapetus. (Credit: NASA/Jet Propulsion Laboratory/Space Science Institute.)

The Smaller Satellites

Hyperion, with an irregular shape (Fig. 1.37) and dimensions of 350×200 km, has a bulk density 1400 kg/m^3. The long axis of the satellite is not aligned toward Saturn as would be expected if gravitational forces had acted on it for a long period of time, which suggests it may have sustained a variety of impacts during its history. Indeed, Hyperion's rotational axis wobbles erratically as it moves along its highly eccentric orbit, and as a result, its orientation in space is virtually impossible to predict. The 3:4 orbital resonance between Titan and Hyperion may contribute to its chaotic rotation as well. Craters ranging from 10 to 120 km across pepper its surface, along with alternating light and dark regions. It seems likely that some of the material (carbonaceous?) eroded off Phoebe by meteoroids may be deposited on Hyperion. Scarp systems as long as 300 km exist on Hyperion, and some rise as high as 30 km above its mean surface level. A substantial amount of H$_2$O-ice probably makes up the interior of Hyperion, but its higher comparative density may point to an admixture of silicate materials. H$_2$O-ice dominates Hyperion's surface as well, but the dark deposits seen here and there are probably

Figure 1.37. This image taken by the Cassini spacecraft reveals the odd shape of Saturn's moon Hyperion and an intriguing variation in brightness across its surface (north is up in the image). (Credit: NASA/Jet Propulsion Laboratory/Space Science Institute.)

carbonaceous in composition. Due to Hyperion's bizarre rotation, its surface is generally uniform, with no preferential accumulation of dark material on the leading hemisphere, as is the case with tidally locked Iapetus.

Synchronized with Tethys in the same orbit, 60° preceding and 60° following it at the L_4 and L_5 Lagrangian points, respectively, are tiny Telesto (~34 × 26 km) and Calypso (~34 × 22 km). Helene (~36 × 30 km) is located 60° ahead of Dione in its orbit. In terms of three-body solar system dynamics, the L_4 and L_5 Lagrangian points are geometric locations where gravitational forces conspire to form stabilized positions for small bodies 60° ahead of and 60° behind a primary attractor like the sun or a major planet (e.g., the Trojan asteroids exist in similar 60° points leading and following the giant Jupiter in its orbit). These co-orbital satellites have cratered surfaces composed primarily of H_2O-ice with an admixture of very dark (carbonaceous?) material. We have already seen earlier how Pandora (~110 × 70 km) and Prometheus (~140 × 80 km) act as shepherd satellites, as shown in Figure 1.23, to keep ring particles within the confines of ring F. We also saw how shepherding moons Janus (~220 × 160 km) and Epimetheus (~140 × 100 km), along with the assistance of Atlas (~50 × 20 km), contribute to the sharpness of the outer edge of ring A. The innermost known satellite of Saturn, the diminutive Pan (~20 km across), is located within Encke's complex and is partially responsible for the existence of the gap. The compositions and surfaces of the six shepherd satellites could be the same as Saturn's co-orbital moons. Not much data exists about the two tiny satellites, S/2004 S1 and S/2004 S2, recently discovered orbiting between Mimas and Enceladus (see Table 1.4). Both are roughly 6 km in diameter and are likely composed of H_2O-ice with some carbonaceous dust mixed in.

Irregular Satellites

Fourteen or so other small moons orbiting Saturn are the irregular satellites. These bodies, sometimes simply referred to as irregulars, orbit Saturn in extremely elliptical paths or they have highly inclined orbits relative to the planet's equatorial plane. Jupiter has a similar retinue of irregular satellites.

Saturnian irregular satellites fall into prograde and retrograde groups, and they are all probably composed mostly of H_2O-ice intermixed with dark carbonaceous material. The seven satellites composing the prograde group reside at quite some distance from Saturn, and thus are weakly bound to the planet. The mechanism of their formation may have involved the close approach of a larger object that was shattered into bits by a collision, or perhaps a parent body was torn apart by a close encounter by Saturn. Some of the resulting fragments possibly wound up in their current orbits with inclinations of ~33° to ~47° to Saturn's equatorial plane (see Table 1.4).

The seven members of the retrograde group of irregulars in Table 1.4 have orbital inclinations ranging from ~153° to ~176°, all relatively close to that of their largest member, Phoebe (inclinations exceeding 90° indicate a retrograde direction of revolution about a primary body, in this case Saturn). Just like the prograde irregulars, the retrograde satellites are most probably composed of H_2O-ice and dark carbonaceous substances. The implication is that the moonlets are small pieces remaining from an impact on Phoebe, although further dynamical studies are necessary. Phoebe, which is 220 km in diameter, is the outermost of Saturn's classical satellites, moving in a retrograde orbit around the planet that is inclined

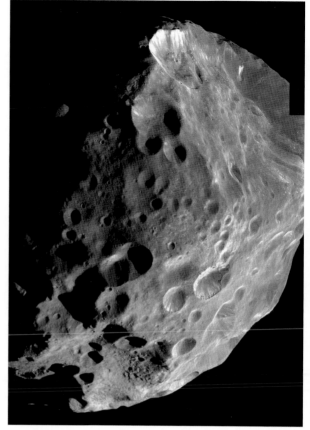

Figure 1.38. Phoebe's violent, cratered past is evident in this mosaic of Cassini images taken on June 11, 2004, with North at the top. Bright material, probably H_2O-ice, appears atop the prominent ridge-like feature in the image as well as around and down the slopes of several craters. Bright streaks are evident on steep slope where loose material possibly slid downhill following the seismic shock of impact events. (Credit: NASA/Jet Propulsion Laboratory/Space Science Institute.)

$174.8°$. It is also the innermost of the retrograde group of irregular satellites, and Phoebe may simply be a chunk of captured interplanetary debris (asteroid or burnt-out comet?) rather than a native satellite of Saturn. Phoebe is roughly spherical in shape, has a rotational period of $\sim10^h$, and its surface is very dark, suggesting that it is a mixture of H_2O-ice and carbonaceous dirt. Close spacecraft flybys of Phoebe reveal a surface covered with impact craters of varying sizes, as well as what appear to be layered deposits of alternating dark and bright material (Fig. 1.38).

Telescopes and Accessories

The three main categories of optical telescopes in use today by observational astronomers are refractors, reflectors, and catadioptrics, although there are numerous design variations in each category. Refracting telescopes use unobstructed optical paths composed entirely of lenses, and two-element achromatic or three-element apochromatic objectives to produce images (Fig. 2.1a). Reflectors are either Newtonians (Fig. 2.1b) or Cassegrains (Fig. 2.1c), and they employ primary and secondary mirrors for image formation. Catadioptrics are compound optical systems that combine lenses and mirrors to create images, and they are either Schmidt–Cassegrains (Fig. 2.1d) or Maksutov–Cassegrains (Fig. 2.1e). While an exhaustive treatment of optical theory and more specialized telescope designs is beyond the scope of this book, this chapter describes some of main features of these instruments, especially how well suited they are for observing Saturn, and discusses some of the more common types of eyepieces employed by planetary observers, as well as a few basic accessories that enhance telescope performance.

Refractors

Because of their simplicity and basic soundness of design, plus the fact that they require only minimal maintenance, refractors have long been the instruments of choice for lunar and planetary observers (Fig. 2.2). The clear aperture of a refractor characteristically produces images of optimum brightness and contrast, since nothing is present in the optical path to reduce light transmission through the objective lens to the eyepiece, and a refractor can often handle more magnification than a reflector or catadioptric of similar aperture. Because a refractor has fewer optical elements than other telescope designs, it adjusts faster to the ambient temperature of the observing environment and thus is ready to use much sooner. To further improve light transmission, most refractors have optical surfaces coated with a thin layer of antireflective magnesium fluoride (MgF_2), while some modern lenses may have premium multicoatings. Also, once the manufacturer sets the optical alignment, a refractor almost never needs any adjustment, even after considerable handling. The closed optical system of a refractor makes it invulnerable to troublesome tube currents that some reflectors suffer from, and only the front surface of the objective lens is exposed to the atmosphere to collect dust and dew, meaning only one lens surface needs to be cleaned. Most refractors have built-

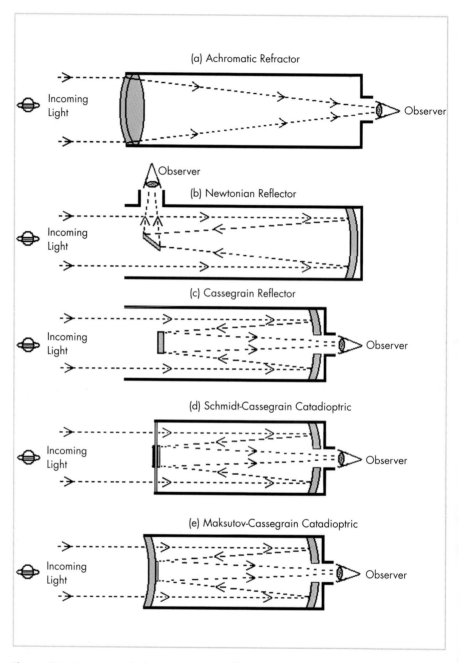

Figure 2.1. Astronomical telescopes are typically available in a variety of designs: (a) achromatic refractor, (b) Newtonian reflector, (c) Cassegrain reflector, (d) Schmidt–Cassegrain Catadioptric, and (e) Maksutov–Cassegrain Catadioptric. (Credit: Julius L. Benton, Jr.)

Figure 2.2. A modern 12.0-cm (4.7-in) achromatic refractor mounted on a sturdy equatorial mounting. Top-quality refractors may be the best all-round instruments for observing planets because of their sharp, high-contrast images. (Credit: Julius L. Benton, Jr.)

in dew caps, but to be effective, they must protrude sufficiently beyond the objective lens.

While the cost of maintaining a refractor is nominal, the initial expense of buying one goes up almost exponentially as the aperture increases in size. It is not unusual, for example, for a refractor to cost four times as much as a Newtonian reflector or Schmidt–Cassegrain of the same aperture. Also, a sizeable refractor often requires a mounting that is considerably more substantial than other telescope designs of identical aperture, so portability diminishes quickly as the objective diameter increases. Part of the reason for this is that refractors used for planetary observing typically have long focal lengths, with focal ratios in the range of f/12 to f/15. For example, a 15.2 cm (6.0 in) refractor with a focal ratio of f/15 has an optical tube nearly 8 feet long!

An achromatic refractor has an objective lens made up of two air-spaced crown and flint optical glass elements, with all air-to-glass surfaces usually coated with a thin layer of MgF_2. All achromats suffer to some degree from longitudinal chromatic aberration, which shows up as a bluish-to-purple halo around extremely brilliant objects. This secondary spectrum becomes more objectionable as the aperture of the instrument increases and its focal ratio decreases. Therefore, long focal ratios of f/12 to f/15 have been the norm for achromatic refractors for many years to help reduce the spurious color. Yet, even though modern achromatic refractors continue to be produced with long focal lengths, a growing number of instruments now have focal ratios of f/6 to f/9 as manufacturers try to maximize portability. As expected, an achromatic refractor of 15.2 cm (6.0 in) aperture and a focal ratio of f/8, for example, has images plagued with the bluish false color, but in actual practice the chromatic aberration is really less objectionable than popular-

Figure 2.3. Newtonian reflectors, like this 20.3-cm (8.0-in) f/6, are among the most popular telescope designs for observing the planets, especially in focal ratios f/6 or greater. (Credit: Mike Karakas, Winnipeg, Ontario, Canada; ALPO Saturn Section.)

ly advertised. Fortunately, the performance of all achromatic refractors, especially those with short focal lengths, can now be measurably improved through the use of selective wavelength blocking filters that suppress the obtrusive violet hues.

Apochromatic refractors have objectives normally composed of three air-spaced optical elements, at least one or more of which may be crafted from extra-low dispersion (ED) glass or fluorite. Apochromats virtually eliminate the objectionable chromatic aberration apparent with achromatic doublets, and all lens surfaces customarily have high transmission multicoatings applied to help optimize image brightness and contrast. Telescope firms can now offer larger and more portable apochromatic refractors with shorter focal lengths without fear of inducing false color, but they are far more expensive than achromats of comparable aperture, especially those with fluorite objectives. While there is no question that apochromats deliver aberration-free images and magnificent planetary detail, the addition of three or more optical elements causes undesirable light loss. Nevertheless, apochromats are in high demand by discriminating observers who have ample funds to spare.

Newtonian and Cassegrain Reflectors

Reflecting telescopes used for observing the planets are generally either Newtonian or Cassegrain optical systems, and they are capable of excellent performance if built to high-quality standards. Newtonians are the simplest of the reflecting telescopes, and they have been traditionally popular among amateur telescope makers (Fig. 2.3). Newtonians use a primary parabolic mirror supported from behind to collect incoming light, which is focused onto a diagonally mounted flat elliptical secondary mirror situated at 45° to the optical path, and in turn directed into an eyepiece at the side of the telescope tube (90° to the optical axis).

The Cassegrain reflector, on the other hand, has a slightly convex secondary mirror in the main optical axis that reflects light back through a central hole in the primary mirror to an eyepiece immediately behind (Fig. 2.4). Primary and secondary mirrors are usually made of glass materials such as Pyrex, with their front optical surfaces coated with a highly reflective layer of aluminum (Al) or special enhanced coatings. Other substrates used in the manufacture of premium Newtonians and Cassegrains include ultralow coefficient of expansion materials like Zerodur, or even quartz in some extreme cases, to preserve the figure of the main mirror (and thus the quality of the image) as the ambient temperature changes during an observing session.

Newtonian reflectors designed mainly for planetary work, with focal ratios between f/6 and f/9, are much more compact than long-focus refractors. For example, while a 15.2 cm (6.0 in) f/8 Newtonian is usually reasonably portable, an f/15 refractor of the same aperture almost certainly requires a semipermanent mount. Even so, as aperture size increases, so does the weight and bulkiness of a Newtonian reflector, so transport from place to place may become a real burden once apertures reach 24.5 cm (10.0 in). Cassegrain reflectors normally have focal ratios ranging from f/10 to f/15, and their design makes shorter tubes possible relative to their aperture for improved mobility. Both Newtonians and Cassegrains require slightly more care and maintenance than do refractors because their alignment must be regularly checked and adjusted to maintain peak optical perform-

Figure 2.4. With focal ratios ranging from f/10 to f/15, Cassegrains are often well suited for planetary work and are quite compact compared with Newtonians and long focus refractors. (Credit: Brian Cudnik, Houston Astronomical Society; ALPO Saturn Section.)

ance, particularly if they are carried around frequently. Restoring alignment in reflectors, however, is quite easy now that inexpensive laser collimating devices are readily available.

Because Newtonian and Cassegrain reflectors have tubes that are open at one end, their optics are susceptible to the effects of moving air currents, dew, dust, and various atmospheric contaminants. Observers sometimes install a small exhaust fan behind the primary mirror as a preventive measure against tube currents, and this also helps retard the formation of dew on the primary mirror. The open tubes of Newtonians and Cassegrains also allow them to reach thermal equilibrium with the environment much quicker than catadioptrics, but not as quickly as refractors.

Classical Newtonians and Cassegrains are obstructed optical systems. The presence of secondary mirrors causes light loss that reduces image brightness, and obstructions in the optical path can adversely influence resolution of delicate, low-contrast planetary detail. This is especially true when the secondary obscuration exceeds about 20% of the diameter of the primary mirror, which is usually near 25% for Cassegrains. Newtonians with focal ratios of f/8 are available with secondary mirrors that produce only 15% obstructions, yet still facilitate a fully illuminated field of view. In this case, light loss as a consequence of the secondary mirror is quite small when compared to the total light-gathering power of the primary mirror. For these reasons, Newtonians often give better results in viewing the planets than do Cassegrains. Indeed, a high-quality Newtonian reflector is a very powerful instrument, fully capable of superb performance in viewing the planets when the optics are kept clean and properly aligned. They have been among the favorite instruments of serious planetary observers for many decades.

Catadioptrics: Schmidt–Cassegrains and Maksutov–Cassegrains

Catadioptric telescopes like the Schmidt–Cassegrain telescope (SCT) and the Maksutov–Cassegrain telescope (MAK) are basically hybrid instruments because

Figure 2.5. A well-equipped 23.5-cm (9.25-in) Schmidt–Cassegrain telescope. Although compact, it is near the upper limit for transport into the field for observing. The SCTs often perform well in viewing planets like Saturn as long as their collimation is correct. (Credit: Jason P. Hatton, Mill Valley, CA; ALPO Saturn Section.)

they utilize a combination of lenses and mirrors to fold the optical path and produce images. The SCTs (Fig. 2.5) form images when light passes through a transparent aspheric corrector plate (lens) onto a spherical primary mirror, which then reflects light back toward a central secondary mirror attached to the corrector plate; in turn, this small secondary mirror sends the light back through a central hole in the primary mirror to an eyepiece behind. Note that since the primary mirror of an SCT is spherical instead of parabolic, a corrector plate is required to remove or "correct" spherical aberration that would otherwise be present.

The MAK (Fig. 2.6) forms images with folded optics much like the SCT, using a spherical primary mirror and a spherical meniscus lens that replaces the aspheric corrector plate used in the SCT. In the MAK, the primary mirror and meniscus lens, since they are both spherical, suffer from spherical aberration, but the manufacturer, to cancel out these optical abnormalities, carefully matches these two elements. Because of the thickness of their meniscus lenses, optical tube assemblies of MAKs are heavier and take longer to "cool down" or reach thermal stability on a given night of observation compared to SCTs of the same aperture.

Figure 2.6. Maksutov–Cassegrains, such as this very fine 18.0-cm (7.1-in) model, are powerful long focal length (f/10 to f/13) instruments in a relatively small package for planetary work. Their collimation usually remains stable even after considerable handling, much like refractors. (Credit: Jan Adelaar; ALPO Saturn Section.)

As with Newtonians, catadioptric telescopes utilize Pyrex or low coefficient of expansion substrates (e.g., Zerodur) to help minimize changes in the figure of their primary mirrors with fluctuating temperatures. Both SCTs and MAKs are closed optical systems just like refractors, so they do not typically experience tube currents, and only one optical surface is exposed to dust and dew. But, unlike refractors, SCTs and MAKs do not usually come with built-in dew caps, so the front corrector plate or meniscus lens will collect moisture very rapidly if an effective dew shield is not used. The SCTs chosen for planetary observation usually have focal ratios of about f/10, while MAKs are mostly f/13 or greater. Because they have their secondary mirror or "spot" vacuum-deposited directly on the meniscus lens, MAKs are far less likely to get out of alignment than SCTs, which have their secondary mirrors mechanically attached through the center of the corrector plate. Adjusting the optics of a SCT can be a tedious, frustrating process, but once established, alignment usually remains intact unless the telescope incurs rough handling.

Because of their folded optical paths, SCTs and MAKs are characteristically very compact and conveniently transportable, much more so that refractors or Newtonians of equal aperture. The SCTs are sometimes referred to as the all-purpose telescope, because they combine in one package many of the favorable attributes of refractors and reflectors, and they also seem to have many more accessories available—especially for photography and CCD imaging—than any other type of telescope. Most observers will agree that a 25.4 cm (10.0 in) SCT is quite manageable when it needs to be moved to and from a remote observing site, and it will fit into a much smaller vehicle as well. Larger SCTs are also much more reasonably priced than refractors or MAKs of the same aperture, while comparable Newtonians are frequently cheaper.

The SCTs and MAKs are obstructed optical systems with secondary mirrors ranging from 28% to 34% of the diameter of their primary mirrors, so some degradation in image contrast and brightness is unavoidable. To partially overcome this problem, manufacturers of top-notch SCTs and MAKs apply special enhanced or "broad-band" coatings to optical elements to improve light transmission of lenses and optimize reflectivity of mirror surfaces. Also, to help increase image contrast and reduce internal reflections, both designs usually have a number of internal light-baffling provisions.

Telescope Mountings

No matter what kind of telescope an observer chooses, it must have an adequate mount that can solidly support the weight of the optical tube assembly and any accessories attached to it, altogether forming a well-balanced setup. It must be sufficiently massive to prevent susceptibility to vibration when a slight breeze is present or when the observer turns the telescope's focusing knob. Once an astronomical object is located and centered in the eyepiece of the telescope, the ideal mounting must be capable of tracking the object as it moves across the night sky without substantial adjustment.

While there are several types of mountings available in the marketplace, instruments intended for serious astronomical observations traditionally come equipped with equatorial mounts. Equatorial mountings have two axes that are set at right angles to each other, the first of which is the right ascension (RA) or polar axis that must be aligned so it is parallel with the rotational axis of the Earth by pointing it toward Polaris, the North Star (for Northern Hemisphere observers). The other is the declination axis (DEC), which is perpendicular to the polar axis, and corresponds with the orientation of the Earth's equator projected against the sky. As the Earth rotates from west to east, the telescope can be moved in right ascension around the polar axis to track an astronomical object and keep it in view, a task easily accomplished by a motorized clock-drive that should be included with the mounting. Equatorial mountings are also normally equipped with manual or electronic slow-motion controls on each axis for making micrometric adjustments in right ascension (east–west motion) and declination (north–south movement). Note that since the declination of stars remains fixed on the celestial sphere, constant tracking in right ascension is all that is required to keep an object in view in the eyepiece. Unfortunately, when it comes to following solar system objects, equatorial mountings will not track precisely in declination, because unlike stars, planets move in the plane of the ecliptic and thus change their declination with time. So, while the mount will track a planet in right ascension, the declination slow-motion control will need to be adjusted to keep it constantly in view. Premium equatorial mounts also usually come outfitted with setting circles on each axis to permit use of right ascension and declination coordinates to find faint objects, although there is really little need for them when the main target of observation is a bright planet like Saturn.

In the last decade or so, computer-controlled telescope mountings have rapidly emerged (sometimes referred to as "go-to" mounts), and they do away with the rigors of having to use setting circles to find faint objects in the sky. Optical encoders are attached to the two axes of such mountings (even altazimuth mounts), which output digital signals as to which direction the telescope is pointing. Preprogrammed computerized handsets allow observers to "dial up" the coordinates of myriad celestial objects, and the telescope will automatically find or "go to" the objects desired. Some highly sophisticated computerized mounts can now even acquire exact Global Positioning System (GPS) coordinates, automatically establish sky directions, and then point the telescope to any one of 150,000 or so objects in the computer's database. For observers who need a convenient, trouble-free way to locate faint deep-sky objects, or perhaps dim solar system bodies such as Uranus, Neptune, Pluto, comets, and asteroids, go-to mountings may be worth the considerable expense. Again, locating brighter planets like Saturn in the sky is seldom an issue, so instead of spending a lot of money on a go-to setup, use the funds to get more aperture!

Although there are quite a few variations of equatorial mountings, the most commonly encountered types are the German equatorial and the fork equatorial mounts. In the German equatorial mount (Fig. 2.7), the declination axis is placed at the end of the polar axis shaft, which in turn is set atop a tripod head that allows proper alignment with the celestial pole (a small refracting telescope is sometimes mounted within the polar axis shaft to facilitate precise positioning). German equatorials usually require longer tripod legs or a higher pier so that the telescope is raised to a convenient observing height, and counterweights are needed at the end of the declination axis opposite the optical tube assembly to exactly balance the weight of the telescope. A well-made German equatorial mounting is capable of truly excellent performance as long as its mass and dimensional characteristics are matched properly to the size of the telescope, and most observers agree that an

Figure 2.7. A sturdy conventional German equatorial mounting with field tripod, slow-motion controls, and a battery-powered clock drive supporting the author's highly portable 12.7-cm (5.0-in) f/13 Maksutov–Cassegrain. (Credit: Julius L. Benton, Jr.)

oversized mount is better than one with marginal stability. Most commercially made refractors, Newtonians, and Cassegrains include German equatorial mountings, although a few catadioptrics utilize them as well.

Fork equatorial mountings are customarily supplied with most commercial SCTs and MAKs because they are easy to use and work efficiently with short-tube catadioptric telescopes (Fig. 2.8). The optical tube assembly is flanked on either side by rigid support arms, the two forming a fork assembly that in turn is secure-

Figure 2.8. An especially solid fork mounting supplied with premium-quality Schmidt–Cassegrain telescopes, outfitted nicely with electronic slow-motion controls, clock drive, and other refinements. Polar axis alignment is accomplished using a special wedge designed for the latitude of the observer. (Credit: Brian Sherrod, Arkansas Sky Observatory.)

ly coupled to a platform that rotates in right ascension. Balance of the telescope occurs in declination between the two fork arms. While these mounts are more compact than German equatorials, they require an adjustable equatorial wedge atop the tripod head or pier to permit accurate polar axis alignment.

Accessories

Star Diagonals, Eyepieces, and Barlow Lenses

Telescopes normally invert and reverse images left to right, and the field orientation seen in any eyepiece will have sky directions (for Northern Hemisphere observers) of south at the top and west to the left. However, as mentioned earlier in this book, when describing planetary phenomena on Saturn, east and west must correspond to true directions on the planet as adopted by the International Astronomical Union (IAU), which are opposite the normal sky directions of east and west. When observing with refractors, Cassegrains, and catadioptrics, the eyepiece may often be situated very low to the ground or in an otherwise uncomfortable position. Star diagonals (Fig. 2.9), which use either a mirror or prism to place images at right angles to the telescope's optical axis, position the eyepiece where observing is more relaxing. Although they may increase observing comfort, star diagonals change the normal (astronomical) inverted and reversed field orien-

Figure 2.9. Prismatic or mirror star diagonals afford more comfortable viewing, but they reduce image brightness and change customary field orientation. (Credit: Julius L. Benton, Jr.)

tation of the eyepiece by flipping the image upright (north is now at the top of the field), which can lead to confusion when describing sky directions or establishing points of reference on images of planets. Great care must be exercised by those who use such devices when recording and referring to directions in their observational notes. Star diagonals, even if they are of the best optical quality, can diminish image brightness and introduce internal reflections. For these reasons, and perhaps at the sacrifice of a little comfort when observing, it is strongly suggested that the use of star diagonals be avoided whenever possible.

Eyepieces (or oculars) function to magnify the image formed by the objective lens or primary mirror of a telescope, so they are an essential part of the overall optical system and should be of equivalent high quality with antireflective coatings (or multicoatings) applied to all lens surfaces. A modest collection of high-quality eyepieces (Fig. 2.10) is certainly preferable to a large repertoire of poor or seldom-

Figure 2.10. A good collection of eyepieces for planetary viewing in a homemade dust-proof wooden case. Observers should avoid using foam-lined cases, because foam deteriorates with time and can damage eyepieces and accessories. (Credit: Julius L. Benton, Jr.)

used ones, while most sophisticated or expensive oculars are not necessarily the best ones for planetary viewing. Eyepieces should be fully compatible with the main optical system, because some that give great performance on f/12 or f/15 instruments might not necessarily perform well with shorter focal length telescopes. So, experimentation with various eyepiece designs on the prime observing instrument is very beneficial.

Most eyepieces today have outside barrel diameters of 3.18 cm (1.25 in) and 5.08 cm (2.0 in) with internal threads that will accept filters. Focusing mechanisms of most commercial telescopes now frequently come with adapters that will accept both eyepiece sizes. There is an almost endless variety of eyepiece designs available on today's market, and focal lengths may typically range from 4 to 50 mm.

The magnification M of a telescope is given by

$$M = \frac{F}{f} \tag{2.1}$$

where F is the focal length of the telescope and f is the focal length of the eyepiece (F and f are expressed in millimeters). Therefore, by Equation 2.1, we can see that a 20 mm eyepiece used on a telescope of 1000 mm focal length will produce a magnification of 50× as follows:

$$M = \frac{1000}{20} = 50\times.$$

Note that shorter focal length eyepieces produce higher powers, and focus adjustment must occur as oculars of different focal lengths are changed on the same telescope. When eyepieces are purchased in sets from the same supplier, they may be parfocal; that is, their barrel assemblies have been designed so that interchanging eyepieces does not require refocusing, which is a major convenience when switching eyepieces at higher magnifications.

The apparent field of view of an ocular, determined by the lens design, is the angular field width seen solely through the eyepiece. Apparent field of view, just like focal length, is a parameter normally listed in a manufacturer's eyepiece specifications, which may range from 60° to 45° for oculars traditionally used in planetary work. A set of eyepieces may all have the same apparent field of view, but different focal lengths. Using Equation 2.2, the true field of view of any eyepiece θ can be determined by dividing the apparent field of view α by the magnification, M as

$$\theta = \frac{\alpha}{M} \tag{2.2}$$

where θ and α are both expressed in degrees. Thus, an eyepiece with an apparent field of view of 50° and a magnification of 25× will have a true field of view of

$$\theta = \frac{50}{25} = 2°.$$

The true field of view of any eyepiece is inversely proportional to the magnification. Accordingly, as observers use higher powers (employing shorter focal length eyepieces), the true field of view will shrink; decreasing magnification (using eyepieces of longer focal length) will expand the true field of view.

Eye relief, ordinarily expressed in millimeters, is the greatest distance that an observer's eye can be positioned from the top lens (eye lens) of an eyepiece and still be able to distinguish the entire true field of view. Longer focal length eyepieces have greater eye relief, while shorter focal length oculars have eye relief that is much more restrictive, especially for observers who must wear glasses while viewing. Rubber eyecups are now provided by many eyepiece suppliers as an added comfort for observers, and they can also help deny entry of extraneous light from nearby artificial illumination.

The diameter of the cone of light emerging from an eyepiece is the exit pupil, d, and can be calculated by using Equation 2.3 as

$$d = \frac{D}{M} \qquad (2.3)$$

where D is the aperture of the objective in centimeters and M is the magnification. So, by Equation 2.3, a 20.3 cm (8.0 in) telescope with a magnification of 100× produces an exit pupil of 2 mm as follows:

$$d = \frac{20.3}{100} = 0.2 \text{ cm or 2 mm.}$$

If the exit pupil exceeds the pupillary diameter of the observer's eye, not all of the light emerging from the eyepiece can reach the retina of the eye, and it is not possible to make use of the full light gathering power of the telescope. Although the diameter of the pupil of the dark-adapted human eye varies from individual to individual and decreases slightly with age, it has a maximum value of about 7 mm. In actual practice, when observing a fairly bright planet like Saturn, the pupillary diameter may be diminished to about 5 mm. Thus, by Equation 2.4, the theoretical minimum magnification M_{min} that can be employed on a telescope and still permit use of the full light grasp of the instrument is

$$M_{min} = \frac{D}{d_{max}} \qquad (2.4)$$

where the D and d_{max} (maximum pupillary diameter) are both expressed in centimeters. Using the maximum pupillary diameter of the eye of 5 mm when observing Saturn, we can see by Equation 2.4 that

$$M_{min} = \frac{D}{0.5} = 2D$$

is about the minimum useful magnification for Saturn. For example, on a 20.3 cm (8.0 in) telescope, this lower magnification threshold would be reached at about

$$M_{min} = \frac{20.3}{0.5} = 40.6 \text{ or } 41×.$$

Obviously, observers will seldom use powers near this lower limit on extended objects like planets since the goal is to obtain optimum image size and brightness, but it is a good theoretical parameter to bear in mind when selecting low-power eyepieces.

Just as there is a theoretical minimum useful magnification for any given aperture, an upper limit also exists. The size of the exit pupil decreases with increasing magnification, and once it reaches a diameter of about 0.75 mm a progressive impairment of vision rapidly begins to set in. This imposes an upper limit on magnifications that can be effectively used on a particular telescope, independent of the prevailing atmospheric conditions and other inconveniences experienced with using eyepieces of extremely short focal length. The theoretical maximum limit of magnification M_{max} that can be employed is

$$M_{max} = \frac{D}{d_{min}} \qquad (2.5)$$

where the D and d_{min} (minimum pupillary diameter) are both expressed in centimeters. Using the smallest acceptable diameter of the exit pupil of 0.75 mm, Equation 2.5 shows that

$$M_{max} = \frac{D}{0.075} = 13D$$

is near the maximum magnification that should be used on a given telescope to avoid degradation in visual acuity. With a 20.3 cm (8.0 in) telescope, therefore, this upper limit would be about

$$M_{max} = \frac{20.3}{0.075} = 270.7 \text{ or } 271\times.$$

Note, however, that the minimum and maximum theoretical limits of magnification discussed here are not absolute limits, and several factors influence these constraints. From extensive experimentation on extended objects like planets, where the best combination of image brightness, contrast, and resolution are so critical for detection of fine detail, maximum magnifications near $25D$ may be closer to the practical limit. But observers should remember that aside from reducing image brightness, higher powers will amplify poor seeing conditions, lessen the field of view, and exaggerate vibrations caused by the mounting or when turning the focusing knob of the telescope.

Some of the most popular types of eyepieces used for planetary observing include Kellners, Orthoscopics, and Plössls (Fig. 2.11). Kellners are composed of three optical elements, with a single eye lens and an achromat for the field lens. They have been consistent high-performers for decades on telescopes with focal ratios exceeding about f/8, producing sharp, generally flat fields of view, and are excellent alternatives to very expensive eyepieces. Since Kellners have comparatively few lenses, they generate high-contrast images with minimal scattered light and internal ghost effects, especially when all lens surfaces are antireflection coated. Kellners have apparent fields of view averaging about 50°, but since their eye relief diminishes with decreasing focal length, they are better suited for low to moderate magnifications. The most common focal lengths of Kellners range from 15 to 25 mm.

Orthoscopic eyepieces are composed of four optical elements—a single eye lens and a triplet field lens—all of which are usually multicoated to eliminate internal

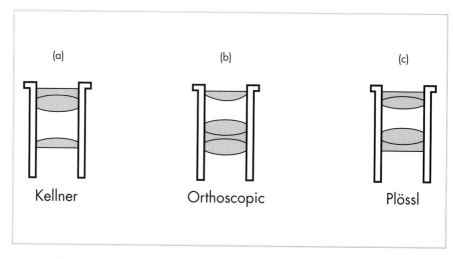

Figure 2.11. The most popular types of eyepieces for planetary observing are three main designs: (a) Kellners, (b) Orthoscopics, and (c) Plössls. (Credit: Julius L. Benton, Jr.)

reflections. They feature decent eye relief, and while they have narrower apparent fields of view (around 45°) than other designs, orthoscopics deliver some of the sharpest planetary images possible. They work best with longer focal length telescopes, and they are often the eyepiece of choice for higher magnifications in focal lengths of 4 through 12.5 mm.

Plössl eyepieces are renowned for their high-contrast, razor-sharp images, apparent fields of view near 55°, and very good eye relief. They utilize the four-element multicoated symmetrical doublet design, and have minimal optical aberrations. Plössls are quite suitable for short and long focal length telescopes, and perform well on virtually all celestial objects, especially planets. Plössls are usually obtainable in focal lengths ranging from 6 through 40 mm, sometimes as parfocal eyepiece sets from a number of premium accessory suppliers.

Observers normally acquire eyepieces in a variety of focal lengths to permit a wide range of magnifications for a particular telescope. Since atmospheric conditions will only occasionally permit observing with the shortest focal length eyepieces, say from 4 to 7 mm, medium-range eyepieces get the lion's share of use. Short focal length oculars also have poor eye relief, and the tiny size of the eye lens in some designs is often objectionable to many observers.

An extremely handy accessory that also makes good economic sense to own is a multicoated, high-quality achromatic Barlow lens (Fig. 2.12), since it will automatically increase the range of powers available with a given assortment of eyepieces. When a Barlow lens is inserted in the focusing mechanism of a telescope between the objective lens or mirror and eyepiece, it functions to amplify the base magnification by 2×, 2.5×, or 3×, and while a Barlow will increase the power of any ocular, its greatest benefit is making higher powers achievable with long focal length eyepieces. Because longer focal length eyepieces have much more eye relief than their short focal length counterparts, observing comfort is markedly improved, especially for those who must wear glasses at the telescope. Barlow lenses are also nearly a prerequisite for telescopes with focal ratios of f/8 or less, since even short focal length eyepieces will not produce adequate higher magnifications alone.

Figure 2.12. Several achromatic Barlow lenses sporting magnifications between 2× and 3×. They conveniently expand the magnification ranges of a given set of eyepieces. (Credit: Julius L. Benton, Jr.)

Finder Telescopes and Reflex Sights

Finder telescopes (Fig. 2.13) are relatively small lower-power achromatic refractors with wide fields of view mounted parallel to and carefully aligned with the main optical system, serving the function of making it easier to locate celestial objects. They usually range in aperture from 3.0 cm (1.2 in) to 7.5 cm (3.0 in) with 6× to 20× cross-hair eyepieces to facilitate precise aiming of the telescope. For them to be truly useful, finders should be of good enough quality to provide pinpoint images across their entire field of view. They should also impart the same field orientation of the image as the main telescope; that is, since the main telescope (used without a star diagonal) provides the normal astronomical inverted and reversed view, so should the finder.

A few equipment manufacturers offer battery-operated reflex sights to supplement, and sometimes even take the place of, conventional finder telescopes

Figure 2.13. A small 6- × 30-mm finder, properly aligned along the optical axis of the main instrument, is sufficient for centering bright planets like Saturn in the telescope's field of view. (Credit: Julius L. Benton, Jr.)

(Fig. 2.14). These extremely useful devices employ a light-emitting diode (LED) to project a small red dot or concentric ring pattern, focused at infinity and adjustable in brightness, onto a clear optical glass window to identify where the main telescope is pointed. Once the reflex sight is initially aligned with the main optical system, it is only necessary to tweak the slow-motion controls of the telescope to superimpose the red dot or concentric rings, which appear suspended in space, on the desired object in the sky. The target should then appear in the center of the field of view of the telescope's eyepiece.

Color Filters and Polarizers

No planetary observer should be without a good set of color filters of precisely known wavelength transmission (Fig. 2.15). Color filters allow observers to significantly enhance their visual perception of fine detail on planets, in particular features that might not otherwise be detected in integrated light (no filter), and they effectively reduce irradiation, making it easier to distinguish boundaries between adjacent light and dark areas. Since planetary surfaces and atmospheres absorb or reflect light in specific wavelengths, it is important to know something about the absorptive and reflective properties of the object being studied so the

Figure 2.14. This small reflex sight is extremely effective in finding objects once its alignment is coincident with the main telescope. (Credit: Jason P. Hatton, Mill Valley, CA; ALPO Saturn Section.)

right filters can be acquired and used effectively. Any given color filter will transmit its dominant color and impede the passage of its complementary color, so for example, orange or reddish Saturnian atmospheric features will appear quite dark in a blue or green filter, yet look bright in a red filter. Likewise, bluish features will be bright in a blue filter and dark in a red filter.

The best-quality color filters for planetary observing are usually manufactured from flat vat-dyed optical glass and are fully antireflection coated to maximize light transmission and eliminate aberrations that degrade images. Ideally, color filters should be mounted in cells that are double-threaded so they can be screwed into standard eyepiece barrels that have diameters of 3.18 cm (1.25 in) and 5.08 cm (2.0 in), while at the same time facilitating "stacking" of filters for cumulative wavelength performance.

Color filters that are probably the easiest to obtain are those based on the industry-standard Eastman-Kodak Wratten series, and each one has a numerical designation that signifies its color. Wratten filters are popular also because of they have precisely known wavelength transmissions, a feature so essential for reliable

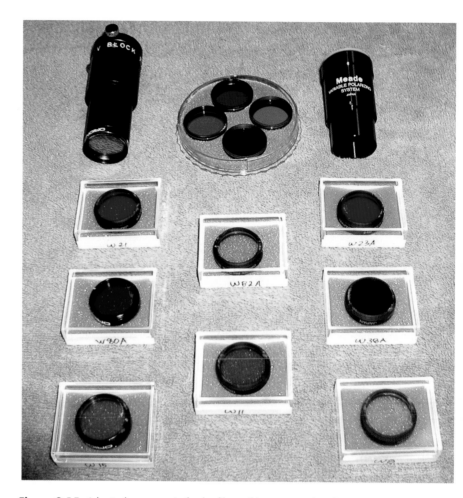

Figure 2.15. A logical assortment of color filters of known wavelength transmissions, along with a good variable-density polarizer, is essential for improving perception of subtle planetary detail. (Credit: Julius L. Benton, Jr.)

colorimetry work, which will be discussed later in this book. Although color filters can be purchased individually, it is generally more convenient and economical to buy them in boxed sets that include the full progression of Wratten colors that span the visible spectrum.

One of the first things that observers will notice when employing color filters is how efficiently they can reduce glare emanating from the brightest zones on Saturn's globe, such as the Equatorial Zone (EZ), and brilliant ring components like the outer portion of ring B. In addition, color filters can selectively transmit certain wavelengths and block others to enhance lower contrast atmospheric features in Saturnian belts and zones (e.g., dusky festoons and bright spots). Table 2.1 lists the characteristics of some of the more popular Wratten color filters employed in planetary work, with their recommended usage when observing Saturn.

In addition to color filters, variable-density polarizing filter systems (Fig. 2.15) significantly improve the visibility of faint atmospheric markings on Saturn, as

Table 2.1. Properties of Selected Wratten Color Filters

Wratten number	Color	Transmission (%)	Dominant wavelength (Å)	Recommended usage for saturn observations
W25	Deep red tricolor	14	6150	Blocks blue and green wavelengths; improves contrast between belts and zones; enhances bluish detail. Most useful with apertures ≥20.3 cm (8.0 in)
W23A	Light red	25	6030	Improves contrast of belts and zones; brings out bluish features; good alternative to W25 filter for smaller apertures
W21	Orange	46	5890	Impedes transmission of blue and green wavelengths; increases contrast and improves visibility of minor detail in belts and zones; good substitute for W25 or W23A filters with smaller apertures
W8	Light yellow	83	5720	Enhances orange and reddish detail in belts.
W15	Deep yellow	67	5790	Helps improve visibility of orange and reddish features in belts.
W11	Yellow-green	62	5500	Very effective in revealing atmospheric detail, especially orange and reddish features.
W58	Green tricolor	24	5400	Blocks red and blue wavelengths; improves visibility of belts and zones, especially bright spots; enhances detail in polar regions.
W82A	Light blue	73	4896	Enhances regions of low contrast; increases visibility of yellow-orange features
W38A	Blue	17	4790	Increases contrast by rejection of red and orange wavelengths; helps reveal subtle detail, including bright spots; enhances contrast of ring components; best used with apertures ≥20.3 cm (8.0 in)
W80A	Medium blue	30	4750	Enhances contrast in belts, zones, and polar regions; markedly improves visibility of fine detail, including ring phenomena; great alternative to W38A for smaller apertures
W47	Blue-violet tricolor	2	4600	Strongly rejects red, yellow, and green wavelengths; very helpful in improving contrast and revealing details in the ring system; best used with apertures ≥20.3 cm (8.0 in)
W30	Light magenta	27	4200 & 6020	Particularly useful for revealing detail in the ring system

well as features in its ring system, by effectively reducing irradiation, and they do so without appreciably affecting the color of the features observed. No serious planetary observer should be without one! Perhaps the most popular type of variable-density polarizer available today comes as an assembly featuring an eyepiece-holder attached to a housing that encloses two premium-grade optical glass polarizing filters, one of which can be rotated relative to the other to vary light transmission from ~1% to 40%. Ideally, the filter housing of the variable-density polarizer should have a threaded barrel at the end opposite the eyepiece adapter that accepts color filters for dual usage.

Choosing the Right Telescope for Observing Saturn

In the foregoing discussion, many of the advantages and disadvantages of various telescope designs, mountings, eyepieces, and other accessories were examined, but the instrument that should ultimately be chosen for viewing Saturn is the one that will be used regularly. The most important consideration when selecting an instrument ought to be its overall optical and mechanical quality, and prospective Saturn observers should simply obtain the best telescope that they can afford. Also, trying to establish an absolute inflexible minimum with respect to aperture for viewing the planet is quite troublesome, because experienced observers have accomplished extraordinary results over the years using surprisingly small telescopes. Almost any optical assistance will show Saturn's spectacular ring system, while major features on the globe are visible with a good 7.5 cm (3.0 in) refractor, including maybe a dusky belt and prominent zone or two near the equator of the planet. Cassini's division is often seen in the rings with such an instrument if the seeing is good. As the eye of the observer becomes accustomed to distinguishing subtle detail on Saturn, refractors and MAKs with apertures of 10.2 cm (4.0 in) or more, and 20.3 cm (8.0 in) or greater Newtonians and SCTs, should suffice for more fundamental studies of variations in belt and zone intensities, as well as enable detailed drawings and CCD imaging of the planet and its ring system. When seeing and transparency conditions are especially favorable, observers can take advantage of the greater image scale, as well as higher resolution and image brightness, of instruments in excess of 25.4 cm (10.0 in) to do advanced, more specialized work.

Since the prime objective is to see the finest detail possible on Saturn and its rings, the telescope one chooses should deliver the very best combination of image brightness, contrast, and resolution. Furthermore, all lenses and mirrors have to be clean, properly aligned, and everything mounted securely so long-term observational work can be performed as comfortably and conveniently as possible. Excellent optics situated atop a stable, functional mounting are of far greater importance than sophisticated go-to electronics.

After acquiring the appropriate instrument and accessories, there is no substitute for quality time spent at the eyepiece actually observing. Regular sessions at the telescope looking at Saturn will train the eye and, in time, will improve one's ability to detect faint detail. At the outset, observers need to experiment with their instrumentation and seek to establish for Saturn the best combination of magnification, image size, brightness, and contrast for the aperture used. It will all pay off in the end.

Factors that Affect Observations

Systematic Observations

Amateur astronomers can observe Saturn regularly, contribute data that are useful to science, and at the same time thoroughly enjoy the experience. Ideally, visual and photographic observations of Saturn and its satellites should be carried out systematically throughout any given apparition, starting early in the observing season after the planet has just emerged from the solar glare after conjunction, then continuing until Saturn again enters the domain of the sun at conjunction. The synodic period (i.e., the time between two successive conjunctions of a planet with the sun) for Saturn is roughly 378^d in length, so in the course of an apparition (which lasts longer by a few days than 1 year on Earth), it will be well placed for observation for about 9 or 10 months (depending on the observer's latitude). Widely spaced observations are of limited value, and the importance of systematic observations by many individuals, all using standardized methods, cannot be stressed strongly enough. Visual observers especially should aspire to achieve the highest possible incidence of objectivity in their data, a feat that can be accomplished when a team of individuals participates in simultaneous observations to monitor variable activity on Saturn. Of course, methods of gathering data by amateur astronomers are rapidly evolving, and it is not unusual to find modern Saturn observers employing electronic devices to supplement routine visual observations. For example, quantitative observations using photoelectric photometers are increasing in number, and the professional-type results some amateurs have been able to achieve using CCD and video imaging techniques is truly remarkable. We will discuss more about photography and CCD imaging later in this book.

Astronomical Seeing

Because all of us, as observers, reside on the surface of the Earth, we must necessarily view all celestial objects, including Saturn, through a thick blanket of air that affects the quality of the images we see in our telescopes. Thus, the state of the atmosphere is a critical factor to consider when we plan and execute observations. Astronomical seeing is the result of a number of very slight differences in the

Astronomical Seeing

refractive index of air from one point to another, and such variations are directly related to density differences, normally associated with temperature gradients, from one location to another. The observed effect of such random atmospheric deviations is an irregular distortion and motion of the image. At one time, the seeing may be "excellent," whereby no gross image fluctuations are noted over a fairly long period of time, while at another instant, the seeing may be "poor," the image appearing as though it is boiling or being seen through a mobile fluid.

It is exceedingly important for planetary observers to establish, as accurately and objectively as possible, the quality of the seeing at the time of observation. When the atmosphere is in a highly turbulent state, it becomes virtually impossible to achieve optimum resolution, and observers are usually forced to wait until conditions are more favorable for useful work. To appraise seeing conditions, therefore, planetary observers typically utilize numerical seeing scales, and among the most popular is the one developed by the Association of Lunar and Planetary Observers (ALPO). This scale ranges from 0.0 (the absolute worst possible seeing) to 10.0 (perfect seeing), where intermediate values are assigned in accordance with an observer's best evaluation of the atmospheric seeing conditions (Table 3.1), preferably in the area of the sky nearest Saturn. Another scale that is quite simple to use is the Antoniadi scale (Table 3.2), whereby observers assign values between 1 (perfect seeing) and 5 (bad seeing) in much the same manner as with the ALPO scale. These scales are completely suitable for most observing situations, despite being somewhat subjective.

More advanced planetary observers might consider using a seeing scale that is more objective and quantitative, one that takes into account the critical relation-

Table 3.1. Association of Lunar and Planetary Observers (ALPO) Seeing Scale

Seeing	Description
0.0–1.0	Worst possible seeing conditions; constant lateral excursion and pulsation of the image; no stability
2.0–3.0	Poor seeing; frequent lateral excursion and pulsation of image; periods of image stability <1.0 second
4.0–5.0	Fair seeing; lateral excursion and pulsation of image apparent; periods of image stability ~1.0 second
6.0–7.0	Good seeing; intermittent excursion and pulsation of image; image stability last a few seconds
8.0–9.0	Excellent seeing; very slight excursion and pulsation of image; image stability last several seconds
10.0	Perfect seeing; no excursion and pulsation of image detected; constant image stability

Table 3.2. Antoniadi Seeing Scale

Seeing	Description
I	Perfect seeing; no image quiver detected
II	Some fluctuations in image; image stability lasts a few seconds
III	Moderate seeing; image quiver much more apparent
IV	Poor seeing; frequent image quiver
V	Worst seeing; constant image quiver

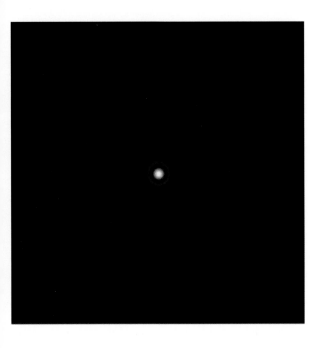

Figure 3.1. Example of an Airy disk as it may appear in near-perfect seeing with in an unobstructed optical system. This image was generated using Aberrator v3.0 developed by Cor Berrevoets. (Credit: Julius L. Benton, Jr.)

ship between atmospheric turbulence and resolution. In reasonably steady air, the image of a star as seen in a quality, properly aligned optical instrument with an unobstructed circular aperture (e.g., a refractor) is a diffraction pattern consisting of a bright diffraction disk, also known as the Airy disk, surrounded by concentric rings of illumination with interspersed dark zones (Fig. 3.1). If a double star is being observed, for example, with two components that are of the same visual magnitude, the instrument employed may or may not be able to distinguish them as two stars. That is, unless the separation of the observed diffraction patterns is at least equal to the radius of the central disk, the two equally bright stellar components will appear as one and unresolved. Increasing magnification will not enhance the capability of the optical system to resolve the stars, mainly because dimensions of the diffraction disks will increase in the same proportion as their separation. Increasing the aperture of the telescope, however, will proportionally reduce the size of the central diffraction discs without altering the separation of two components. Thus, aperture is the key factor affecting resolution of the two stars.

The angular radius θ of the Airy disk in radians is given by

$$\theta = \frac{1.22\lambda}{D} \tag{3.1}$$

where λ is the wavelength and D is the aperture of the instrument (both values expressed in centimeters). For our purposes here, it is necessary to convert radians to arc seconds ($''$) by multiplying the value of θ by 206,265 because there are 206,265$''$ per radian. Thus, Equation 3.1 may be rewritten as

$$\theta = \frac{1.22\lambda}{D}(206,265).$$

Since the maximum visual spectral sensitivity of the human eye is attained at about 5500 Å (expressed in centimeters this is 5.5×10^{-5}) under photopic conditions. For a 1.0 cm aperture, the value of θ is

$$\theta = \left[\frac{1.22(5.5 \times 10^{-5})}{1.0} \right] 206,265 = 13.84''.$$

Rayleigh's theoretical limit of resolution, R, of any given aperture, assuming stars of identical visual magnitude against a dark background sky, is defined by Equation 3.2 as

$$R = \frac{13.84''}{D} \tag{3.2}$$

where D is expressed in centimeters as before. Thus, Rayleigh's criterion tells us theoretically whether or not we will be able to distinguish between the two stars observed, and it should be obvious that this quantitative expression is also dependent on wavelength as well as aperture.

Anatomically, the iris controls the amount of light that enters the eye through its pupillary aperture, which has a variable diameter anywhere from 2.0 to 8.0 mm depending on the light intensity (field brightness) and in some instances on the elasticity of the iris. In addition, visual acuity is normally at its keenest within the pupillary aperture range of 2.0 to 6.0 mm. Below this range, the resolution of two points of light against a dark background is diffraction limited, and beyond it aberrations set the limit on the resolution of the eye.

If the assumption is made that the pupillary diameter of the eye is 5.0 mm, appropriate for most observing situations in planetary astronomy when the background of the sky is considered dark (i.e., no daylight observations), the resolution of the eye determined by using by Equation 3.2 is

$$R = \frac{13.84''}{D_e} = \frac{13.84''}{0.5 \text{ cm}} = 27.7''$$

under the stated conditions, where D_e is the diameter of the pupil of the eye expressed in centimeters. While the theoretical resolution of the eye as determined above is 27.7″, the limit of resolution is probably much nearer 70.0″ or even 140.0″ in actual practice.

The minimum magnification that may be employed on a given aperture, D (expressed in centimeters), whereby the eye is expected to resolve everything the telescope resolves, is given by Equation 3.3 as

$$M_{min} = \frac{\theta_e}{13.84'' / D} \tag{3.3}$$

where θ_e is the resolution of the eye in arc seconds as before and D is the aperture of the telescope in centimeters. If we adopt as the minimum resolution of the eye the very conservative value of 140.0″ as mentioned earlier, we may employ Equation 3.3 accordingly:

$$M_{min} = \frac{140.0''}{13.84'' / D} = \frac{140.0'' \, D}{13.84''} = 10 \, D.$$

So, when magnifications of at least 10 D are used on a given instrument, for all practical purposes the only factors limiting the telescope's resolution are its aperture and the prevailing seeing conditions. Because seeing is rarely perfect, the theoretical limits of resolution of a specific instrument as determined by Rayleigh's theoretical criterion are seldom attained.

It is possible to express, therefore, the resolution of a certain instrument on a given night of observation as the effective aperture, D', of the telescope, using Equation 3.4 as

$$D' = eD \qquad (3.4)$$

where e is the efficiency of the instrument.

There are a number of ways to determine the efficiency of a particular telescope on any night of observation. Before using the instrument, however, it is rather important that the telescope be permitted to overcome thermal shock by adjusting adequately to the ambient temperature. The easiest method to ascertain the efficiency of any telescope is to derive the ideal resolution for a 2.54 cm (1.0 in) aperture, r, which is a personal constant. The telescope used most frequently in planetary observation must be "stopped down" in aperture to precisely 2.54 cm (1.0 in). Next, the observer must select a number of double stars of about the same visual magnitude and with angular separations ranging from 4.0″ to 6.0″ (separations should be constant). The closest double star that can be barely resolved with a 2.54 cm (1.0-in) aperture provides the personal constant, r, previously noted. This procedure should be attempted on a night of exceptional seeing, and once this constant is established, it is sufficiently stable and may not need checking for several months.

Now, during each observing session, a second quantity is measured, known as the actual resolution for a 2.54 cm (1.0-in) aperture, r'. The closest double star that can be resolved using full aperture of the instrument in question is noted, and the separation distance of the double star is subsequently multiplied by the full aperture of the telescope use (in centimeters), which yields the value, r', as

$$r' = s(D) \qquad (3.5)$$

where s is the separation of the double star observed. The efficiency of the telescope is given by Equation 3.6 where

$$e = \frac{r}{r'} \qquad (3.6)$$

and the effective aperture, D', is determined by Equation 3.7 as

$$D' = \frac{rD}{r'}. \qquad (3.7)$$

As an example, let's suppose that the value of constant r was established as 5.45″ using a 15.2 cm (6.0 in) refractor "stopped down" to 2.54 cm (1.0 in) under absolutely superb seeing conditions on a given date (as appraised using one of the more subjective scales discussed earlier in this chapter). Now, during a subsequent observing session, employing the same instrument at full aperture, presume that it was only possible to resolve a pair of double stars of equal magnitude against a dark sky with a separation of 1.50″. So, using Equation 3.5,

$$r' = s(D) = 1.50''(15.2 \text{ cm}) = 22.8''$$

and the efficiency of the instrument, therefore, by Equation 3.6, is

$$e = \frac{r}{r'} = \frac{5.45''}{22.8''} = 0.239 \text{ or } 23.9\%.$$

The effective aperture for the given night of observation is, by Equation 3.7,

$$D' = \frac{rD}{r'} = \frac{5.45''(15.2 \text{ cm})}{22.8''} = 3.63 \text{ cm},$$

suggesting that the resolution realized with the 15.2 cm (6.0 in) refractor was nearer that of a 3.63 cm (1.43 in) telescope!

Once an observer has achieved some skill in measuring the effective aperture using this simple method, he may be able to estimate accurately the value of D' from previous experience by closely examining the stability and sharpness of the image. Table 3.3 lists about a dozen or so sample double stars that have relatively constant angular separations that observers may use as approximate standards of resolution (more comprehensive double-star tables for this purpose are available in the appropriate literature as well as on the Internet). It is worthwhile to bear in mind, though, that stars differing somewhat in brightness are considerably more troublesome to resolve than stars having equal magnitudes and the same separation.

In essence, then, the above method is perhaps useful as a means for quantitatively evaluating the state of the atmosphere in relation to resolution. Again, whenever possible, the effective aperture, D', should be determined by experienced observers instead of making subjective estimates of the seeing. It must be stressed that stars used for determining r or r' should be as close to the zenith as possible to avoid detrimental effects of atmospheric dispersion and differential refraction, which is usually very pronounced at low altitudes. The stars should also be as near Saturn as possible in the sky (or at least at the same altitude), and viewing the planet when it is high in the sky is always advised.

Experienced planetary observers are generally familiar with meteorological conditions that yield the best seeing conditions for their location, and sometimes the clearest nights, so essential for deep-sky observing, are not necessarily the best for planetary work. For example, the author has repeatedly encountered excellent seeing conditions on warm, hazy nights when the relative humidity in coastal Georgia has been 90% or higher. In general, locations that experience calm evenings, where air masses are homogeneous and undergo minimal nightly fluctuations in temperature and barometric pressure, are more desirable for good planetary work.

Topography also plays a role in determining the quality of astronomical seeing. Observers should keep away from the vicinity of valleys or low-lying terrain into which cold air flows and where heavy fog and mists are likely to form. High eleva-

Table 3.3. Selected double stars for quantitative estimates of seeing

Star Name	HD	SAO	Cons	RA hh	RA mm	RA ss	DEC °	DEC '	DEC "	m_v (total)	m_1	m_2	Separation (")
ADS 161	895	73823	And	00	13	23.9	+26	59	14	6.3	6.5	8.0	0.1
66 Psc	5267	92145	Psc	00	54	35.2	+19	11	18	5.80	6.2	6.9	0.4
φ And	6811	36972	And	01	09	30.1	+47	14	31	4.25	4.8	5.4	0.5
48 Cas	12111	4554	Cas	02	01	57.3	+70	54	26	4.48	4.8	6.5	0.3
α Psc	12447	110291	Psc	02	02	02.7	+02	45	49	3.79	4.3	5.2	1.5
ADS 2612	21903	24111	Cam	03	35	00.7	+60	02	28	6.46	6.8	7.8	0.8
52 Ori	38710	113150	Ori	05	48	00.2	+06	27	15	5.27	6.0	6.1	1.4
12 Lyn	48250	25939	Lyn	06	46	14.1	+59	26	30	4.87	5.4	6.0	1.8
ADS 5447	49059	96111	Gem	06	47	23.4	+18	11	36	6.20	6.8	7.0	0.5
ζ Cnc	68257	97645	Cnc	08	12	12.6	+17	38	52	4.67	5.7	6.0	0.9
ADS6762	70340	135804	Hya	08	21	20.1	−01	36	09	6.50	7.0	7.3	0.6
57 Cnc	75959	61125	Cnc	08	54	14.6	+30	34	45	5.39	6.1	6.6	1.4
φ UMa	85235	27408	UMa	09	52	06.3	+54	03	51	4.59	5.0	5.5	0.4
ADS7704	88987	99032	Leo	10	16	16.0	+17	44	24	7.30	7.2	7.4	1.1
ι Leo	99028	99587	Leo	11	23	55.4	+10	31	45	3.94	4.1	4.7	1.0
μ Dra	154906	30239	Dra	17	05	19.5	+54	28	13	4.92	5.5	5.5	1.8
τ Oph	164764	142050	Oph	18	03	04.8	−08	10	49	4.79	5.2	5.9	1.7
73 Oph	166233	123187	Oph	18	09	33.7	+03	59	36	5.73	6.1	7.0	0.5
ε¹	173582	67309	Lyr	18	44	20.1	+39	40	15	4.67	6.0	5.1	2.6
ε²	173607	67315	Lyr	18	44	22.7	+39	36	46	5.10	5.1	5.4	2.2
λ Cyg	198183	70505	Cyg	20	47	24.3	+36	29	27	4.53	4.7	6.1	0.7
2 Equ	200256	126482	Equ	21	02	12.2	+07	10	47	7.4	7.7	7.4	2.8
52 Peg	217232	108307	Peg	22	59	11.7	+11	43	44	5.75	6.1	7.4	0.7
72 Peg	221673	73341	Peg	23	33	57.0	+31	19	31	4.98	6.0	6.0	0.5

HD, star's designation as listed in the *Henry Draper Catalogue*; **SAO**, the star's designation as listed in the *Smithsonian Astrophysical Observatory Star Catalogue*; **Cons**, constellation; **RA**, right Ascension; **DEC**, declination; m_v **(total)**, combined visual magnitudes of both stars; m_1 and m_2, visual magnitude of each star; **Separation**, distance between the two stars in arc seconds (").

Astronomical Seeing

tions, such as mountainous plateaus, are mostly well above dust and air pollution and are usually wonderful sites from which to view the planets, although it is advisable to avoid the crest of a hill where moving air currents and windy conditions create turbulence. Some of the worst areas to set up a telescope for observing are concrete driveways or asphalt parking lots that have been exposed to sunlight all day. Heat waves radiated throughout much of the night will ruin what might otherwise be decent seeing. Terrain covered by grass or low shrubbery facilitates better seeing conditions because vegetation reduces thermal overheating of the surroundings during the daytime.

Transparency

In addition to making a determination of the seeing conditions during an observing session, the transparency of the atmosphere should also be evaluated by estimating as precisely as possible the visual magnitude (usually denoted by m_v) of the faintest star just barely perceptible to the observer's dark-adapted, unaided eye. This method has been in use for decades and remains the adopted technique for assessing sky transparency by most planetary observers.

The initial step in estimating atmospheric transparency requires that an observer accurately determine his personal correlation coefficient, C. This is easily achieved by using a reliable star atlas and finding the magnitude of the faintest star (to the nearest 0.25 visual magnitude), in this case denoted by m_z, that can be seen by the dark-adapted, unaided eye in the zenith on a clear, dark, moonless night well away from artificial illumination. Using Equation 3.8, the observer's personal correlation coefficient, C, is

$$C = 6.0 - m_z. \tag{3.8}$$

Next, the observer must accurately estimate, again to the nearest 0.25 visual magnitude, the faintest star, denoted by m_p, just discernible to the dark-adapted, unaided eye in the immediate vicinity of Saturn. In making this determination, it is assumed that there is no twilight or moonlight, and essentially no artificial light interference, and the atmospheric transparency, T_r, may then be computed using Equation 3.9 as

$$T_r = m_p + C. \tag{3.9}$$

Once again, the quantitative determination of atmospheric transparency as described above presumes ideal circumstances, where little or no extraneous light is present in the region of the sky nearest Saturn and likewise mostly absent from the zenith. In situations where there is considerable light interference from a nearby city, from the moon, or due to twilight, then the best that one can do is estimate the faintest star, to the nearest whole magnitude, that would be visible independent of the scattered light. This is often facilitated by reference to some other characteristic of the sky, such as its depth of blueness, clarity at twilight, etc. It is important also to understand that sky transparency is a logarithmic expression of the light transmission properties of the atmosphere, not a function of extraneous or scattered light. Thus, T_r values will be affected by fog, mists, and haze.

Resolution, Image Brightness, and Contrast Perception

In our discussion of astronomical seeing, the theoretical limits of resolution as defined by the Rayleigh criterion do not hold if the two stars being observed are of unequal brightness (as with double stars of different visual magnitudes or different spectral type), nor does it apply thoroughly to resolution of detail on extended objects like planets. In actual practice, experienced observers with especially keen vision, under excellent seeing conditions, are able to resolve double stars of the same magnitude with angular separations below Rayleigh's threshold with a given aperture. This empirical resolution criterion, or Dawes's limit, denoted by R_d, was established long ago by visual observations of double stars, and is defined by Equation 3.10 as

$$R_d = \frac{11.58''}{D} \tag{3.10}$$

where D is expressed in centimeters. For example, assuming perfect seeing conditions, Rayleigh's criterion predicts that the theoretical limit of resolution of a 15.2 cm (6.0 in) telescope is, using Equation 3.2,

$$R = \frac{13.84''}{D} = \frac{13.84''}{15.2 \text{ cm}} = 0.91''$$

while by Equation 3.10, Dawes's limit for the same instrument is

$$R_d = \frac{11.58''}{D} = \frac{11.58''}{15.2 \text{ cm}} = 0.76''$$

Just as with Rayleigh's criterion, Dawes's limit is essentially applicable only to stars of the same visual magnitude and spectral type when observed against a dark background sky. Table 3.4 lists comparative Rayleigh and Dawes resolution limits for various apertures.

Since planets are extended objects, their images are made up of overlapping Airy disks and diffraction rings formed by light reflected from every point on their visible surfaces, and Rayleigh's and Dawes' criteria cannot effectively be applied to the resolution of planetary detail. Skilled observers are routinely able to detect, in good seeing, delicate features on the surfaces and in the atmospheres of planets far below the theoretical resolution limits of any given telescope. For example, Cassini's division, which is only about 0.5" to 0.6" wide at Saturn's ring ansae, can be seen under good seeing conditions with only a 6.0 cm (2.4 in) aperture refractor, which has a theoretical Rayleigh threshold of 2.27" and empirical Dawes's limit of 1.90".

Luminance, L, is the amount of visible light emanating by emission, transmission, or reflection from a surface in a given direction. The standard unit of lumi-

Table 3.4. Theoretical limits of resolution by telescope aperture

Telescope aperture		Rayleigh's criterion	Dawes empirical limit
cm	in	(″)	(″)
1.00	0.39	13.84	11.58
2.54	1.00	5.45	4.56
5.08	2.00	2.72	2.28
6.10	2.40	2.27	1.90
7.62	3.00	1.82	1.52
8.89	3.50	1.56	1.30
10.16	4.00	1.36	1.14
12.70	5.00	1.09	0.91
15.24	6.00	0.91	0.76
17.78	7.00	0.78	0.65
19.05	7.50	0.73	0.61
20.32	8.00	0.68	0.57
22.86	9.00	0.61	0.51
23.50	9.25	0.59	0.49
25.40	10.00	0.54	0.46
27.94	11.00	0.50	0.41
30.48	12.00	0.45	0.38
31.75	12.50	0.44	0.36
33.02	13.00	0.42	0.35
35.56	14.00	0.39	0.33
38.10	15.00	0.36	0.30
40.64	16.00	0.34	0.29
43.18	17.00	0.32	0.27
45.72	18.00	0.30	0.25
48.26	19.00	0.29	0.24
50.80	20.00	0.27	0.23

nance is the candela per square meter (cd/m^2), and accordingly, the surface brightness, S, of any planet may be conveniently expressed in cd/m^2. The visual geometric albedo, p_v, of Saturn is 0.47, which is the percentage of incident light that is reflected by the planet in the direction of the observer. The true surface brightness of Saturn's globe has been determined accurately and is ~180 cd/m^2. The true surface brightness value for Saturn, as well as that of the rest of the planets, changes with the phase angle, i, or the planetocentric angle between the Earth and the sun, and with the distance of the planet from the sun. This is particularly the case for the inferior planets like Mercury and Venus, for which the numerical value of the phase angle, i, becomes significantly greater than 0°, and as a consequence, the surface brightness of the planet in question will be appreciably less than its visual geometric albedo. A phase angle of 0° occurs when the hemisphere facing the Earth is fully illuminated (at opposition for a superior planet or at superior conjunction for an inferior planet), while 180° occurs when the facing hemisphere of the planet is completely dark (as is the case only at inferior conjunction for inferior planets). Inferior planets and the moon exhibit all possible phase angles, while superior planets like Saturn show only a small phase angle. Because the numerical value of the phase angle for Saturn does not vary much beyond 0°, and since its perihelion and aphelion distances are not substantially dissimilar, the true surface brightness of ~180 cd/m^2 for the planet remains fairly constant.

Before reflected sunlight from Saturn can enter the eye of the observer, it is to some extent diluted by absorption in the atmosphere, in the telescope's optical system, and in any filter(s) that may be used while observing the planet. The total transmission of an unobstructed optical system, like a refractor, independent of any filters used, may be upward of 85% or so, but may be little more than 80% for Newtonians and as low as 60% to 70% for catadioptrics. Enhanced reflective coatings on mirror surfaces and high-transmission multicoatings on lenses improve matters somewhat, but light loss still occurs before the image reaches the retina of the observer's eye. With respect to filters, the percentage of light transmission for a number of Wratten filters appears in chapter 2, Table 2.1, and it is a factor contributing to the overall transmission characteristics of the optical system when the filters are used. As a cumulative result of all these factors, Saturn always has an apparent surface brightness lower than its true measured value of $S = 180$ cd/m^2. The aperture of the telescope and the magnification employed also play a critical role, whereby the image of Saturn is brighter by the ratio of the area of the telescope's aperture to the area of the pupil of the eye, and is fainter by the square of the magnification.

Contrast is defined as the fractional difference in brightness between two objects. For example, if two areas on the globe of Saturn have a true surface brightness of S_1 (a bright zone such as the EZ) and S_2 (a dusky belt like the South Equatorial Belt (SEB)), where $S_1 \geq S_2$ as measured in cd/m^2, the true contrast, c, between the two areas on the globe is

$$c = \frac{S_1 - S_2}{S_1}. \tag{3.11}$$

If c is 0.0, then S_1 must equal S_2, and the two areas are of equal brightness, and when $c = 1.0$, S_2 must be 0.0 (perceptibly black). Now, as may be observed through a given telescope, the apparent contrast, c', between the two Saturnian surface features is defined by

$$c' = \frac{S'_1 - S'_2}{S'_1} \tag{3.12}$$

where S'_1 and S'_2 denote apparent surface brightness in cd/m^2, and $S'_1 \geq S'_2$.

Assuming that the two atmospheric features on Saturn are fairly large in comparison with the limiting resolution, c' will only be noticeably different from c when scattered light is present in the image. In our example, if the true surface brightness of S_1 (EZ) and S_2 (SEB) is taken to be 180 and 90 cd/m^2, respectively, the true contrast, c, between the two areas is derived by Equation 3.11 as

$$c = \frac{180 - 90}{180} = \frac{90}{180} = 0.50 \text{ o } 50\%.$$

As mentioned earlier, light absorption in the atmosphere, telescope, and any filters used will produce a proportionate diminution in the light transmission for the two areas, but in the absence of scattered light, remember that c is not appreciably different from c'. Thus, the apparent contrast between S'_1 and S'_2 is also 50%. If scattered light—say, 30 cd/m^2 from the bright EZ—is added to the adjacent dusky SEB, then the apparent contrast c' between the two areas by Equation 3.12 becomes

$$c = \frac{150-120}{150} = \frac{30}{150} = 0.20 \text{ or } 20\%.$$

and contrast suffers considerably between the EZ and SEB due to the presence of a small amount of scattered light.

In general, it has been shown by experiment that relative contrasts are more noticeable if the apparent surface brightness of a planet lies between 300 and 3000 cd/m^2, while optimum contrast perception is attained at brightness levels near 1000 cd/m^2. To have an apparent surface brightness as high as 1000 cd/m^2, it would be necessary for a significant reduction in magnification to occur. For Saturn, this is not possible, because even an apparent surface brightness of 100 cd/m^2 is virtually impossible to achieve (recall that the true surface brightness of the planet is only 180 cd/m^2), taking into consideration the light absorption that occurs due to the instrument, atmosphere, and filters employed. For magnifications less than about 10 D, as was emphasized earlier, the eye is usually not capable of resolving all that the optical system can resolve. Experimentally, the lowest possible light level for reasonably good contrast perception has been shown to be ~10 cd/m^2, so for Saturn, observers must use much lower magnifications than 10 D. Therefore, larger apertures are necessary when seeing and transparency conditions permit.

So that the observer may achieve optimum contrast perception, and thus be able to detect discrete phenomena in the atmosphere of Saturn, it is essential that the image be large and bright. The contrast sensitivity of the human eye may be evaluated in terms of the minimum perceptible difference in brightness between two contiguous areas, yet it is extremely difficult to obtain contrast perception that approaches the theoretical visual threshold in planetary studies. If the magnification is increased in an attempt to get a bigger image, the planet will frequently be too dim; also, if the magnification is reduced to remedy this problem (i.e., in an effort to achieve a brighter image), the result is often an image so small that nothing can be seen to advantage on the planet's globe. So, a happy medium is sought so that both requirements of large image size and brightness can be satisfied.

Image size is a factor that plays an even more significant role in contrast perception than does the level of surface brightness. Seasoned observers using high magnifications on comparatively dim planets like Saturn often disagree that lower powers will improve perception of delicate contrasts. There is a strong dependence, in truth, of contrast perception on the angular dimensions of various atmospheric features on Saturn (e.g., belts and zones), and it turns out that more is usually gained in discrimination of delicate contrasts by a moderate increase in magnification than will be lost by the resulting diminution in apparent surface brightness.

Again, by experimentation, it has been possible to determine the approximate magnifications for use on the planet Saturn that will yield optimum contrast perception, assuming excellent observing circumstances and reasonably clean optics. Calculations have been made for those features that have an angular diameter nearly twice the Rayleigh criterion for resolution, as well as for features that comprise a significant fraction of the planetary dimensions. For Saturn's finest atmospheric details and elusive features in the rings, very high magnifications ranging from about 350× to 500× are necessary to see them to advantage. It is obvious that large apertures are mandatory for good image characteristics at these powers. Larger surface features may be detected with low to moderate magnifications, in the range extending from about 200× to 400×.

It is worth mentioning that, given any magnification in excess of 10 D (where it is remembered that D is expressed in centimeters), the resolution of surface detail is at a maximum value when the contrast perception is also at the optimum level. This is because Rayleigh's criterion is valid only for sources of infinite contrast, while the surface details on Saturn frequently differ in brightness only to a small degree, so the contrast has to be good for detail to be seen. The magnification ranges noted here can be considered as the upper limit on occasions when the conditions for viewing Saturn are absolutely perfect, and lower magnifications are more useful under average conditions as long as they are not significantly less than 10 D. Since Saturn is generally observed against a dark sky background, there may be some psychophysical influences on contrast perception. Also, turbulent seeing conditions often spread out and initiate boundaries that are ill-defined for Saturn's atmospheric features, so lower magnifications than those suggested might be required to sharpen peripheral areas and improve perception of delicate, often elusive detail.

Perception of Color

Under normal circumstances of illumination, the visual wavelength sensitivity of the human eye ranges from 3900 Å (violet) to 7000 Å (deep red) along the electromagnetic spectrum, with maximum sensitivity occurring at about 5500 Å. Two kinds of photo-receptive cells are located in the retina of the eye, known as the rods and the cones. The rods are active when illumination levels are extremely low (below about 0.034 cd/m^2), and they are responsible for dark-adapted night vision, also referred to as scotopic vision. They contain a pigment called "visual purple" that has utmost sensitivity at 5100 Å in the green region of the electromagnetic spectrum, and scotopic vision is monochromatic (i.e., there is no color vision).

Also comprising the retina are three types of cones that contain photosensitive pigments having maximum wavelength responses of 4450 Å (blue), 5350 Å (green), and 5725 Å (red) and function in the illumination range above 3.4 cd/m^2. The wavelength sensitivities of the retinal cones overlap, producing a composite sensory response that forms the basis for color vision under daylight conditions throughout the visual spectrum, also referred to as photopic vision.

For intermediate ranges of illumination between the 0.034 and 3.4 cd/m^2 thresholds, when light levels are low but it is not completely dark, a combination of photopic and scotopic vision occurs, known as mesopic vision.

The surface brightness of a planet is of far greater significance in visual color acuity than in contrast perception. As noted above, the color-sensitive cones are not functional at light levels below 0.034 cd/m^2, and if the image of Saturn is exceedingly dim, it is probable that a number of chromatic illusions will interfere with the observation, rendering the results highly questionable. At a brightness level of 0.034 cd/m^2, where the transition occurs from photopic to scotopic vision, the Purkinje effect causes objects to look bluer than normal. Above the brightness range where the Purkinje effect occurs—between 0.50 and 50 cd/m^2—another complex illusion plays a role, known as the Bezold–Brüke phenomenon. In this case, red, yellow, green, and blue colors associated with Saturnian belts or zones appear as normal, but yellow-green and orange hues on the globe take on a more yellowish appearance, while blue-green and violet colors look substantially bluer. All this takes place as the brightness of the image diminishes, and it is apparent that reddish or greenish colors are subtracted somehow. As planetary surface brightness levels become very great, all colors show a marked reduction in their saturation.

Direct color estimates are dependent on the angular extent of the Saturnian feature being observed, and especially for blue, green, and violet hues, the color becomes more saturated as dimensions of the feature under study become appreciable. Atmospheric phenomena with the smallest apparent angular size are affected in such a way that colors of violet or yellow-green show up as gray, while other colors appear bluish-green or reddish-orange.

On a more practical note, it is recommended that observers frequently shift their eye from one point on the globe of Saturn to another adjacent location, chiefly because keeping the eye essentially fixed on any one spot for a long period of time tends to produce fading in color and contrast of nearby areas of differing color and contrast.

Simultaneous contrasts, produced when one color is superimposed on a background of a different hue, present numerous problems for planetary observers. Neutral or unsaturated colors, which are superimposed on more saturated hue backgrounds, frequently assume the complementary color of the background. Gray, for instance, on a reddish background appears greenish! From an exhaustive investigation of these phenomena, it seems clear that induced contrast colorations are more or less insensitive to fluctuations in surface brightness, and the more saturated the surrounding hue, the more obvious the contrast-induced color. Also, the contrast-generated effects become more noticeable the longer one stares at a particular region on Saturn, the smaller the size of the feature, and the more indistinct the boundaries (as may result from poor seeing or excessive magnification).

Absolute color estimates by visual means are made less subjective and more standardized through the use of color reference charts that are available from a variety of photographic or art suppliers when viewing Saturn.

A Word About Simultaneous Observations

Simultaneous observations, which involve two or more observers working independently but viewing Saturn on the same date and at the same time, employing analogous equipment and methods, substantially reduce the number of observational variables and the level of subjectivity inherent in visual work. Including CCD and webcam imaging in this effort only helps strengthen the value of resulting comparative observational data (see Chapter 9). The importance of concurrent long-term systematic work by dedicated Saturn observers using standardized methods cannot be emphasized enough.

Visual Impressions of Saturn's Globe and Ring System

As discussed in Chapter 1, Saturn has a mean synodic period of 378^d, the interval between successive conjunctions of the planet with the sun. Therefore, an apparition of Saturn is only just a little longer than one terrestrial year. Saturn's annual eastward motion relative to the background stars amounts to ~12°, so it usually remains in one constellation for an extended period.

To the unaided eye, Saturn is at least a first-magnitude object, displaying a very distinct yellowish hue. The optimum time to observe the planet from mid-northern latitudes on Earth occurs at opposition, when it appears on the celestial meridian at midnight and rides relatively high in the sky where atmospheric turbulence, notoriously common near the horizon, is minimal. When Saturn is at opposition and closest to the Earth, the equatorial diameter of the globe subtends nearly 20.5″ and its majestic rings typically span about 47.0″.

The tilt angle between Saturn's rotational axis and pole of its orbit, or obliquity, is 26.7°. Therefore, its axis of rotation always has essentially the same orientation in space, and relative to the sun and to our line of sight (because Earth is an inferior planet), the degree of inclination of the globe and rings changes with time. Saturn, therefore, exhibits seasons much like the Earth. From our terrestrial vantage point, over the course of one Saturnian year (29.5 Earth-years), the value of B, or Saturnicentric latitude of the Earth referred to the ring plane, varies between 0° to ±26.7°. When B is positive (+) the northern hemisphere of Saturn's globe and north face of the ring system can be viewed to advantage from Earth (especially near maximum inclination of +26.7°). Likewise, when B is negative (−) optimum views are possible of the planet's southern hemisphere and south ring surface. When B is 0.0°, the rings are situated edge-on to our line of sight, and equal portions of the north and south halves of Saturn's globe are visible.

Saturn's globe does not often exhibit the same frequency and conspicuousness of atmospheric detail and activity that is so common on Jupiter. As a result, novice observers are often disappointed with their first visual impressions of Saturn, especially at lower powers and with smaller apertures. With time, however, patient observers will gradually see more and more detail at the threshold of vision during moments of good seeing and optimal image brightness and contrast.

This chapter dicusses the types of features and phenomena commonly reported on Saturn's globe and in the rings, derived from impressions of persistent, experienced observers of the planet after many observing seasons. Interspersed here and there are several drawings and images from the archives of the Association of Lunar Planetary Observers (ALPO) Saturn Section to help illustrate visual impressions of globe and ring features and associated phenomena; unless otherwise noted, south is always toward the top of each drawing or image, and east is to the left in the International Astronomical Union (IAU) sense in the normal inverted astronomical view. Later chapters discuss the most important Saturn observing programs and contemporary methodology for recording useful data (including opportunities for amateur–professional collaboration). Here is a handy digest of the endeavors serious amateur astronomers are currently involved in that readers can refer to as we discuss visual impressions of Saturn's globe and rings:

- Visual and numerical relative intensity estimates (visual photometry) of belts, zones, and ring components
- Full-disk drawings and sectional sketches of the globe and ring system using standard observing forms
- Central meridian (CM) transit timings of details in belts and zones on Saturn's globe
- Filar micrometer measurements or estimates of the latitude of belts and zones on Saturn's globe
- Colorimetry and absolute color estimates of globe and ring features
- Observation of "intensity minima" in the rings in addition to studies of Cassini's, Encke's, and Keeler's divisions
- Systematic color filter observations of the bicolored aspect of Saturn's rings, as well as monitoring of suspected azimuthal ring brightness asymmetries
- Observations of stellar occultations by Saturn's globe and rings
- Specialized studies of Saturn during edgewise ring orientations in addition to routine studies
- Visual observations, charged coupled device (CCD) imaging, and magnitude estimates of Saturn's satellites
- Multicolor photometry and spectroscopy of Titan to confirm a suspected rotational light curve variation of 7% at 940 nm
- Regular imaging of Saturn using webcams, digital and video cameras, and CCDs, ideally coincident with visual observations as part of an overall simultaneous observing program
- A comprehensive simultaneous observing program (all-inclusive)

Telescopic Appearance of Various Regions of the Globe

In considering the belts and zones traditionally visible on Saturn's globe, it is customary to progress pole to pole from the southern hemisphere through the northern hemisphere as seen in the normally inverted view of an astronomical telescope (refer to the diagram in Figure 1.3 in chapter 1 and compare it with the excellent drawing in Figure 4.1).

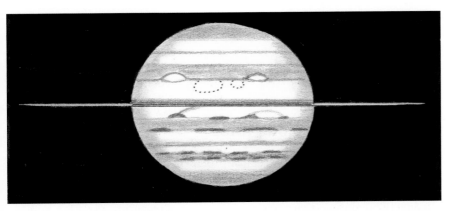

Figure 4.1. Drawing contributed to the ALPO Saturn Section by Philip Budine of Binghamton, New York, using a 25.4-cm (10.0-in) Newtonian at 350× in excellent viewing conditions on October 21, 1966, at 02:50 Universal time (UT). Saturn's rings appear edgewise to our line of sight, where equal portions of Saturn's two hemispheres are visible. Compare this drawing with the nomenclature presented in Figure 1.3. (Credit: Philip Budine; ALPO Saturn Section.)

Analogous belts and zones exist in each hemisphere of the globe, and it is standard practice for observers to compare corresponding latitudinal features whenever possible. Relative studies of similar atmospheric features and phenomena in the same hemisphere of the globe are also meaningful. When making drawings and descriptive reports, it is important to remember to document any relative similarities or differences between regions and features.

The Southern Hemisphere of Saturn

South Polar Region (SPR)

The SPR usually appears yellowish-gray during many observing seasons, and it sometimes exhibits small variations in brightness. Periodically visual observers may detect a distinct dusky gray south polar cap (SPC) that is recurrently duller than its surroundings, but not always. A dark-gray south polar belt (SPB) encircling the SPR may occasionally be present, and this curvilinear feature can now and then look darker that many of the other belts of Saturn's southern hemisphere.

South South Temperate Zone (SSTeZ)

The pale yellowish-white SSTeZ is rarely a very conspicuous zone on Saturn, and it is perhaps easier to see with apertures exceeding 31.8 cm (12.5 in). Visual observers occasionally describe slight variations in the brightness of the SSTeZ, including rare sightings of discrete atmospheric phenomena, such as small short-lived white ovals.

South South Temperate Belt (SSTeB)

Larger apertures in excess of 31.8 cm (12.5 in) are usually required to see the narrow light grayish-white SSTeB as it extends across the globe of Saturn from one limb to the other. Visual observers seldom report activity in the SSTeB.

South Temperate Zone (STeZ)

The yellowish-white SteZ remains mostly uniform in brightness throughout most apparitions, sometimes mimicking the brightness of the nearby south tropical zone (STrZ), and may occasionally approach the conspicuousness of the equatorial zone (EZs) during some observing seasons. Comparing the STeZ with its complement in the northern hemisphere, the NTeZ, both zones are time and again identical in brightness throughout apparitions when both can be seen simultaneously (i.e., when the inclination of the rings to our line of sight is relatively small). Atmospheric phenomena do not seem to arise frequently in the STeZ, but transient features may appear at any time to watchful observers, particularly at ring inclinations in excess of −20°.

South Temperate Belt (STeB)

The dull grayish STeB is one of the more commonly seen belts on Saturn's globe, especially with apertures of 20.3 cm (8.0 in), extending uninterrupted across the globe from one limb to the other. The STeB is often a little darker than its northern hemisphere counterpart, the NTeB, during observing seasons when one can be compared with the other. Small dark spots within the roughly linear STeB may occasionally be visible to keen-eyed observers.

South Tropical Zone (STrZ)

Visual observers routinely report the yellowish-white STrZ with apertures of 10.2 cm (4.0 in) and greater, and it normally remains stable in brightness between apparitions. The STrZ may approach the EZs in overall conspicuousness, and during many observing seasons the STrZ and STeZ are often similar in appearance. Small ephemeral white spots emerge within the STrZ during some apparitions, persisting for only a few hours to several days (Fig. 4.2).

South Equatorial Belt (SEB)

Considered as a singular feature, the grayish-brown SEB is perhaps the most distinct belt in the southern hemisphere of Saturn, showing little fluctuation in brightness. During most apparitions, however, visual observers usually describe it as differentiated into SEBn and SEBs components (just like the NEB), with an intervening dull yellowish-gray south equatorial belt zone (SEBZ). While the SEB, considered as a whole, is typically the darkest and most eye-catching belt in the

Figure 4.2. This drawing by Carlos Hernandez of Miami, Florida, with a 22.8-cm (9.0-in) Maksutov at 248× in good viewing conditions at 06:40 UT on November 9, 2004, shows a small white spot in the STrZ bordering the south equatorial belts (SEBs.) (Credit: Carlos Hernandez; ALPO Saturn Section.)

southern hemisphere of Saturn, the SEBn is commonly slightly darker than the adjacent SEBs. Visual observers with different apertures regularly detect diffuse dark spots and dusky projections emanating from the northern border of the SEBn, extending into the EZs during many apparitions. Figure 4.3 is a good example of the kind of discrete phenomena sometimes visible along the SEBn neighboring the EZ when viewing conditions are above average in moderate apertures. Of course, atmospheric phenomena are normally easier to see with larger telescopes. Most dusky features in the SEBn or SEBs are short-lived and do not

Figure 4.3. This beautiful sketch by Sol Robbins, observing from Fair Lawn, New Jersey, using a 15.2-cm (6.0-in) refractor at 350X in excellent viewing conditions at 03:00 to 03:25 UT on February 5, 2003, shows several undulations along the SEBn adjacent to the EZs. (Credit: Sol Robbins; ALPO Saturn Section.)

remain visible for several rotations of Saturn to facilitate systematic recurring CM transit timings. On the other hand, in rare instances when the same feature is visible for successive rotations, it is extremely important to accurately record CM transits (see Chapter 7).

Saturn's Southern Hemisphere: A Special Note

As this book goes to press, it is perhaps worthwhile to emphasize that Saturn reached perihelion on July 26, 2003, which occurs every 29.5 terrestrial years (one Saturnian year). Some investigators believe that a slight increase in atmospheric activity in the southern hemisphere of Saturn may be a response to the planet's seasonal insolation cycle, although measurements in the past imply a relatively slow thermal response to solar heating at Saturn's distance from the Sun of 9.0 astronomical units (AU) at perihelion. Nevertheless, observers are encouraged to keep Saturn's southern hemisphere (Fig. 4.4) under close surveillance during forthcoming apparitions following the planet's perihelion passage, since a lag in the planet's atmospheric thermal response may possibly imitate a similar pattern we experience on Earth, where the warmest days do not occur on the first day of summer, but a few weeks later.

Figure 4.4. This image of Saturn's southern hemisphere taken by Damian Peach of Norfolk, United Kingdom, using a 23.5-cm (9.25-in) Schmidt–Cassegrain and a Philips ToUcam at 20:07 UT on February 19, 2003, shows numerous details that visual observers using larger apertures may be able to glimpse when seeing conditions permit. It is always meaningful to compare visual impressions of Saturn with quality CCD images such as this. (Credit: Damian Peach; ALPO Saturn Section.)

Equatorial Zone (EZ)

When Saturn's rings are near the time of edgewise orientation, as in 1996 and again in 2009, the pale yellowish-white EZ has two halves, the EZn (region of the EZ between where the rings cross the globe and the NEB) and the EZs (portion of the EZ between where the rings cross the globe and the SEB). The EZ is virtually always the most brilliant zone on Saturn's globe, closely approximating the brightness of ring B, and considerable atmospheric activity may occur in this region. For those periods when the rings are close to, but not exactly edgewise to, our line of sight, observers may see the EZn or EZs through portions of the very tenuous Ring E where it crosses in front of Saturn's globe, which may contribute to perceived disparities in brightness between the two halves of the EZ.

Very large and prominent white spots emerge periodically within the EZ, the most recent of which was the Great White Spot of 1990 (Fig. 4.5), which was so

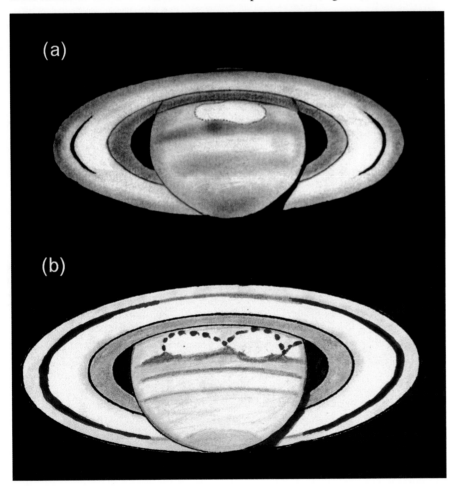

Figure 4.5. Truly massive white spots emerge in Saturn's EZ at intervals of roughly 57 years, like the major outburst of October 1990: (a) by Jean Dragesco in France using a 35.6-cm (14.0-in) SCT at 244× and (b) by Philip Budine with a 10.2-cm (4.0-in) refractor at 250× from Binghamton, New York. (Credit: Jean Dragesco; Phil Budine; ALPO Saturn Section.)

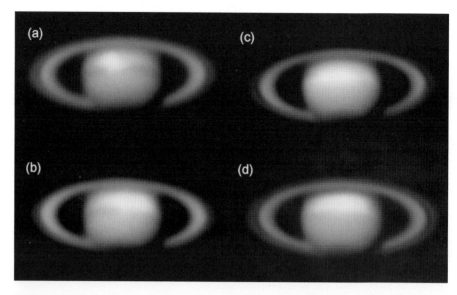

Figure 4.6. The Great White Spot of 1990 evolved from a centralized eruption in Saturn's EZ then gradually spread out longitudinally over time in this sequence of photographs by Isao Miyazaki of Okinawa, Japan, using a 40.6-cm (16.0-in) Newtonian: (a) October 2, 09:39 UT; (b) October 10, 11:00 UT; (c) October 17, 10:50 UT; and (d) October 22, 10:04 UT. (Credit: Isao Miyazaki; ALPO Saturn Section.)

bright that even the smallest apertures permitted excellent views of the dazzling atmospheric disturbance.

Convection processes uplift ammonia (NH_3) into Saturn's outer frigid atmosphere, where bright ice clouds condense and become visible as brilliant white spots. Within a few weeks, the 1990 storm elongated and developed into a bright streak, eventually occupying virtually all of the visible EZ. Such rapid evolution apparently involved an early convective eruption that signaled the beginning of the white spot flare-up, an expansion longitudinally in an east-to-west direction, followed by a maturation phase where the bright spot diffused and eventually encircled the planet's equatorial region (Fig. 4.6). The Great White Spot of 1990 occurred during the Saturnian summer season in the northern hemisphere of the planet, just like analogous outbursts in 1876 and 1933. Complex mechanisms appear to occur with increasing or decreasing insolation, but whether or not a seasonal effect contributes to the formation of white spot eruptions in the EZ is unclear. It is important to realize, too, that Saturn has an internal heat source that alone might be adequate to account for such occurrences. Comparative analytical studies of long-term systematic observations of the renowned white spots of 1876, 1933, and 1990 by amateur astronomers confirm that they emerged at a regular interval of 57 years, a little less than two Saturnian years. Furthermore, smaller EZ bright spots were visible in 1990 for a few months after the main eruption, mimicking phenomena seen in 1876 and 1933, suggesting that these massive storms had not completely dissipated. Indeed, because there seems to be a 57-year interval for such spectacular white spot disturbances, the youngest Saturn observers reading this book can prepare for the next anticipated occurrence in 2047! Although the aforementioned major EZ white spot outbreaks garnered a lot attention by visual

Figure 4.7. This drawing by Carlos Hernandez of Miami, Florida, depicts the narrow, sometimes elusive EB. He used a 22.8-cm (9.0-in) Maksutov at 343× to sketch Saturn in good viewing conditions on November 26, 2003, at 07:00 UT. (Credit: Carlos Hernandez; ALPO Saturn Section.)

observers, other moderately large white spots can arise sporadically within 10° on either side of Saturn's equator. Many of these persist long enough for worthwhile CM transit timings before dissipating, and any elevated brightness of the EZ can probably be attributed mostly to the emergence of such diffuse white spots.

The equatorial band (EB), usually described as a dark gray, ill-defined linear feature running across Saturn's globe, is occasionally visible in larger apertures in favorable viewing conditions. In the drawing presented in Figure 4.7, the narrow EB is clearly visible.

The Northern Hemisphere of Saturn

North Equatorial Belt (NEB)

The grayish-brown NEB may often be seen differentiated into NEBn and NEBs components, especially in telescopes larger than 20.3 cm (8.0 in), where "n" refers to the north component and "s" to the south component, separated by a somewhat diffuse yellowish-gray NEBZ (north equatorial belt zone), but the NEB just as likely may be visible as a singular feature, too. When the NEBn and NEBs are both seen, the NEBs is frequently the darker of the two. In smaller telescopes, however, the most common aspect of the NEB is as a single feature, only rarely subdivided into components. Ordinarily one of the dullest zones on Saturn, the NEBZ is easier to see because it is sandwiched between the darker NEBn and NEBs. In moments of better viewing, observers may regularly distinguish ill-defined, transient dusky projections and condensations within the NEBn and NEBs, but, unfortunately, few of these features ordinarily persist long enough to facilitate good CM transit timings. During many observing seasons, the NEB is often the most conspicuous and darkest belt of Saturn's northern hemisphere, showing barely perceptible brightness changes with time. The NEB is often nearly equal in brightness compared with its southern counterpart, the SEB.

North Tropical Zone (NTrZ)

The yellowish-white NTrZ is frequently second only to the equatorial zone (EZ) in being the brightest zone on Saturn's globe, and it shows very slight intensity changes over several apparitions. Sometimes an occasional bright area or festoon may appear as it crosses the globe.

North Temperate Belt (NTeB)

The ordinarily light-grayish NTeB is frequently a very ill-defined feature on Saturn's globe, as a rule only barely discernible from its immediate surroundings. On occasions when it is visible, the NTeB is quite diffuse as it extends across the globe from limb to limb. Possibly contributing to the poor contrast of the NTeB with adjacent zones on Saturn is the comparatively similar brightness of these features.

North Temperate Zone (NTeZ)

The dull yellowish-white NTeZ periodically exhibits small fluctuations in brightness from apparition to apparition, and it is customary for observers to report short-lived diffuse light and dark features associated with the NTeZ, all usually near the threshold of vision. When the tilt of Saturn's rings permits visibility of both features during an apparition, the NTeZ and STeZ often exhibit nearly the same brightness.

North North Temperate Belt (NNTeB)

It is not unusual for there to be a few isolated sightings of a narrow grayish NNTeB during any observing season, but the best chances to see this elusive belt take place with apertures in excess of 31.8 cm (12.5 in).

North North Temperate Zone (NNTeZ)

During any given apparition, observers may catch a glimpse of a dull yellowish-white NNTeZ, but sightings of this feature are uncommon with apertures less than 31.8 cm (12.5 in). Fluctuations in the overall brightness of this feature may occur sporadically.

North Polar Region (NPR)

The yellowish-gray NPR typically maintains uniformity in overall appearance from one apparition to the next, but subtle variations in brightness do occur from time to time. A dusky yellowish-gray north polar cap (NPC), often just a little darker than its immediate environs, may appear now and again during an observing season near the extreme north limb of Saturn. Sometimes a narrow grayish north polar belt (NPB) encircles the NPR in good seeing.

Telescopic Appearance of Saturn's Rings

In a previous chapter we discussed the three major, or classical, ring components making up the ring system of Saturn, at least for most visual observers on Earth. Without its ring system, Saturn would be little more than a dimmer, largely unimpressive, and tranquil replica of the giant Jupiter. Beside their aesthetic qualities, the rings are responsible for much of Saturn's brightness, and there is nothing comparable to them, at least in terms of prominence and extent, anywhere in the solar system.

The three major or classical ring components, as depicted in Figure 1.3 in chapter 1, are ring A (the usually seen outermost component), ring B, (the central, broader ring), and ring C (the inner dusky Crape ring). A dark gap called Cassini's division separates rings A and B, and it is visible with ease in a 6.0-cm (2.4-in) refractor when the rings are fully open to our line of sight. Halfway out from the globe of Saturn in ring A is another division, although less well defined as Cassini's, called Encke's division. It is readily visible with 31.8-cm (12.5-in) telescopes in good viewing conditions, and some observers have described it as being multiple on occasion, resulting in the designation "Encke's complex." Ring C is by far the faintest of the ring components described, but it is sometimes visible with a 7.5-cm (3.0-in) telescope at the ansae when viewing and transparency conditions are excellent. In Figure 4.8 Saturn's major ring components are plainly visible in the sketch.

As noted in Chapter 1, in addition to the three major ring components described above, there is a very elusive innermost ring D, located just internal to ring C. Some individuals have suggested that they have seen ring D visually in front of the globe on occasion (or perhaps the shadow of ring D), but confirmation of these sightings is lacking. External to ring A is a very tenuous, broad ring E, the initial observations of which date back to the 1907–1908 apparition when the rings were edgewise to the Earth. Ring E was confirmed by spacecraft at the same time that ring F (just outside ring A) and the tenuous ring G (which extends outward from ring F) were discovered in recent years. Both components F and G are within the confines of ring E, however, so ring E remains the outermost of the known ring components. Rings F and G have very little importance to visual observers on Earth except perhaps when occultations of stars occur by the rings. Even then, it is very uncertain whether amplitude data caused by ring F or G is visually measurable from Earth.

Aside from the clearly defined "gaps" of Cassini and Encke, observers have reported for many years various "intensity minima" in the rings that interrupt the otherwise continuous ring components. These fine, elusive divisions exist in great

Figure 4.8. In this very fine drawing by Sol Robbins of Fair Lawn, New Jersey, on October 8, 2003, at 08:00 to 08:15 UT using a 15.2-cm (6.0-in) refractor at 400×, all three of the major ring components are represented, as well as Cassini's division and Encke's complex, and several "intensity minima" in ring B. (Credit: Sol Robbins; ALPO Saturn Section.)

multitudes, as confirmed by spacecraft, so the rings are hardly unbroken across their breadth (Fig. 4.8). Some of the intensity minima so conspicuous in spacecraft photographs may be within reach of very large apertures from Earth when viewing conditions cooperate. The characterization and positioning of observed intensity minima is of great importance, particularly because they vary both in prominence and in position with time. Because of the actual myriad abundance of these features, there is little hope that an observer on Earth will discover any new intensity minima in the rings.

The ALPO Saturn Section archives show that there are about 10 recognized intensity minima, in addition to Cassini's and Encke's divisions, that are recognizable from Earth. Of course, larger telescopes in excess of 40.6 cm (16.0 in) are required to view these with any predictable success in good viewing conditions, and measurements of their positions relative to the component they occur in, as well as to each other, are always useful. To provide a means for accurate and easy identification of ring divisions and more subtle intensity minima, a convenient system exists for denotation of such features. Observers simply assign a capital letter and a number to the division or intensity minimum seen. The letter denotes the ring component the gap is located in, and the number indicates the relative position of the division as a fraction of the distance outward from the globe into the ring. Encke's division (or "complex" as it is sometimes called), for example, is designated E5 because it is about 50%, or about halfway, out from the globe of Saturn in ring A, while Keeler's gap is denoted as A8 because it is about 80% of the way out in ring A. Likewise, Cassini's division is A0 or B10 because it is exactly at the boundary between rings A and B. This method is also very convenient for assigning nomenclature to fainter, often ephemeral, intensity minima in each ring component, such as B1, B3, and B5.

In discussing the visual impressions of the rings of Saturn, the usual convention (adopted here) is to begin nearest the globe and proceed outward across the expanse of the rings, and then consider some special observational conditions and phenomena associated with the rings.

Ring C

Ring C, most often visible at the ansae, exhibits a consistent grayish-black appearance throughout most apparitions of Saturn, with only subtle fluctuations in brightness. Usually, faint or narrow features in the rings, like ring C, are more evident and typically appear darker at greater ring inclinations. Ring C is relatively easy to see with 7.5-cm (3.0-in) telescopes at such times. The Crape band, or ring C as it crosses in front of the globe, is often quite recognizable in 7.5-cm (3.0-in) to 10.2-cm (4.0-in) apertures. The Crape band is usually uniform in intensity and appears dark yellowish-gray to dark gray in appearance. Except when they are near the plane of the rings, the Saturnicentric latitudes of the sun and Earth conspire to bring about the partial coincidence of the Crape band with the shadow of ring C on the globe; thus the Crape band may appear darker than it actually is.

Ring B

The outer third of ring B is the adopted standard of reference for the ALPO Saturn Intensity Scale, with an assigned value of 8.0 (visual photometry is discussed later in this book). This region of ring B is always the brightest portion of the rings, and it is typically the brightest feature on both the globe and the rings. The outer third of ring B in most observing seasons appears consistently white and remains rather stable in brightness. On a few rare occasions, however, the brightness of the EZ may closely approach that of the outer third of ring B, and major white spot outbursts

Figure 4.9. Several small ring "spokes" are visible at the W and E ansae in ring B in this drawing by Sol Robbins, Fair Lawn, New Jersey, on November 14, 2004, at 07:40 to 08:04 UT with a 15.2-cm (6.0-in) refractor at 400×. (Credit: Sol Robbins; ALPO Saturn Section.)

within the EZ may easily rival in brilliance this portion of the ring. The generally dimmer, yellowish-white inner two thirds of ring B usually does not fluctuate much in brightness throughout and between observing seasons. Visual observers also occasionally suspect dark-gray intensity minima at B1, B2, and B5 positions within the inner third of ring B (apparent in Fig. 4.8), and vague dusky "spokes" infrequently appear in this ring component near the ansae at greater ring inclinations (Fig. 4.9).

Cassini's Division (A0 or B10)

Cassini's division (A0 or B10) is quite easy to detect at the ansae with 6.0-cm (2.4-in) apertures when the rings are inclined substantially toward the Earth and approaching the maximum value of B of $\pm26.7°$. At such times, with larger instruments and good viewing conditions, this grayish-black gap is typically visible around the circumference of Saturn's ring system. As the rings reach inclinations less than $\pm10°$, in a progression toward edgewise orientation, Cassini's division is usually still quite apparent at the ansae. This gap may not appear as black at less significant ring tilts with smaller instruments, but any divergence from a completely black intensity for Cassini's division is simply a result of poor viewing conditions, scattered light, or inadequate aperture.

Ring A

The yellowish-white ring A, considered as a whole, usually ranks third behind the outer and inner thirds of ring B in brightness during most observing seasons. When the rings are sufficiently inclined toward Earth, with values of $\pm15.0°$ or more, visual observers report pale yellowish-white outer and inner halves of ring A, with the outer half of ring A often marginally brighter than the inner half. The dark grayish Encke's division (A5) is occasionally visible at the ansae in excellent viewing conditions with instruments exceeding 10.3 cm (8.0 in) in aperture, sometimes described as "Encke's complex" because of its apparent multiplicity to most visual observers. Observers also report sporadic sightings of the steel-gray Keeler's division (A8) further out in ring A with 31.8-cm (12.5-in) telescopes or larger when atmospheric viewing conditions are better than average. Delicate azimuthal brightness and color oscillations in ring A also emerge from time to time, and they seem to arise when light is scattered by higher density particle clumps orbiting in ring A.

Bicolored Aspect of Saturn's Rings

The bicolored aspect of the rings is an observed variance in coloration between the east and west ansae (IAU system) when they are systematically compared with alternating W47 (Wratten 47), W38, or W80A (all blue filters) and W25 or W23A (red filters). Visual observers report this curious phenomenon during most apparitions, and it is sometimes quite apparent when similar effects are absent from the east and west limbs of Saturn's globe; that is, if it is visible on both the globe and

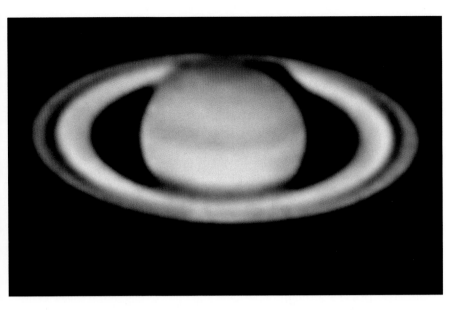

Figure 4.10. This digital image of Saturn taken by Clay Sherrod of Little Rock, Arkansas, on January 20, 2002, at 00:50 UT with a 30.8-cm (12.0-in) SCT purportedly shows the bicolored aspect in ring A at the E and W ansae of Saturn's rings. Saturation of hues is exaggerated in this image to help illustrate the appearance of this very subtle effect. Notice also that azimuthal brightness variations in ring A may be apparent in this image. (Credit: Clay Sherrod; ALPO Saturn Section.)

rings, the effect would undoubtedly be due to atmospheric dispersion. A few rare photographs and digital images over the years allegedly show the phenomenon as well (Fig. 4.10). Possible causes of the bicolored aspect include atmospheric dispersion, vignetting in the telescope, a short-lived solitary confinement of particles according to size within the rings, as well as an illusion due to motion of the eye away from the telescope's optical axis at high magnifications and small exit pupil, coupled with the use of specific color filters. A final, definitive explanation for the bicolored aspect of the rings awaits further research.

Appearance of Saturn's Rings at Edgewise Orientations

Throughout Saturn's sidereal revolution period of 29.5^y, the intersection of the orbit of the Earth and the plane of the ring system takes place only twice, at intervals of 13.75^y and 15.75^y. The two periods are of unequal length because of the ellipticity of Saturn's orbit around the sun, and the rings are edgewise to our line of sight at these times ($B = 0.0°$). Astronomically speaking, such events are quite rare and particularly noteworthy. During the 13.75^y period, the south face of the rings and the southern hemisphere of the globe are inclined toward the Earth, and Saturn reaches perihelion during this interval. In the slightly longer 15.75^y period, Saturn passes through aphelion, and the north face of the rings and the northern hemisphere of the globe are visible to observers on Earth. The last edgewise presentation of the rings took place around the time of opposition in 1995–1996, with Saturn well placed for viewing much of the night by numerous observers world-wide. Subsequent edge-on apparitions occur in 2009 and 2025, neither of which will be particularly noteworthy because Saturn will be too close to conjunction with the sun, so observers will have to wait until 2038–2039 to see another favorable edgewise orientation when the planet is near opposition.

At edgewise orientations of Saturn's rings, observers attempt to determine how close to their theoretical edge-on position the rings are visible in various apertures. The apparent disappearance of the ring system, which may occur several times during a short interval, is due to several geometric circumstances. First, the Earth may lie in the plane of the rings so that only their edge is visible to observers, and since the rings are quite thin, they are temporarily lost to even the largest telescopes. Second, the sun may lie in the ring plane so that only their edge is illuminated. Lastly, the sun and Earth may lie on opposite sides of the ring plane, so that observers on Earth see regions illuminated only by light that is passing directly through the rings (i.e., forward scattering).

Larger instruments, in the range of 30.8 cm (12.5 in) to 41.0 cm (16.0 in), are necessary to facilitate observations of the sunlit side of the ring system up to within a few days or even hours of the dates and times of theoretical edgewise presentation. With regard to the dark side of the rings, visibility may elude observers for several days or even weeks prior to and following edgewise presentations (Fig. 4.11 and 4.12). Thus, there is a general asymmetry with respect to the extent, appearance, and brightness of the rings of Saturn at such times. For instance, especially near the precise edgewise orientation of the ring system, nonuniformities in

Figure 4.11. This drawing of Saturn's edgewise rings on September 2, 1995, at 05:30 UT was made by the author using a 15.2-cm (6.0-in) apochromatic refractor at 300× in viewing conditions. Notice the difference in the extent of the rings on either side of Saturn's globe, a phenomenon commonly reported by visual observers near edge-on orientations. (Credit: Julius L. Benton, Jr.; ALPO Saturn Section.)

Figure 4.12. Color photograph of Saturn's edge-on rings taken from Nagoya, Japan, by Toshihiko Ikemura at 16:19 UT on August 19, 1995, using a 21.0-cm (8.3-in) Newtonian. (Credit: Toshihiko Ikemura; ALPO Saturn Section.)

brightness sometimes appear as condensations of light along the otherwise dark ring surface.

A meaningful endeavor is to estimate the numerical relative intensity of the illuminated and dark ring surfaces throughout the apparition at various distances from the globe of Saturn. Because the outer third of ring B (the normal intensity reference standard at 8.0) is not clearly visible when the rings are edgewise, observers must resort to using an alternative feature on Saturn's globe, usually the EZ, as the reference point for visual numerical relative intensity estimates. Fortunately, the EZ is usually stable in brightness in the absence of white spot outbreaks, and has the arbitrary value of 7.0 for comparison purposes at these times.

The intensity of the ring system at different positions is apparently proportional to particle density; so light passing through the rings from the illuminated surface, therefore, is complementary. In other words, the intensity of the dark side is opposite that of the sunlit side. Thus, the outer third of ring B, instead of being the brightest portion of the rings under sunlight conditions, is the darkest area. Ring A, according to the same scenario, is significantly brighter, and ring C may appear as the brightest component of all! Complications obviously result from light reflected onto the rings by the Saturn's globe, but this illumination should essentially vary as the inverse square of the distance from the planet.

The very elusive, vast, and dusky ring E external to ring A elicits some mixed impressions during edgewise apparitions, times when ring E is presumably easier to see from Earth in good viewing conditions with larger instruments. Although ring E is definitely present, as confirmed by spacecraft, controversy persists in some circles as to the visibility of the component in Earth-based telescopes.

Anytime the dark side of the ring system faces Earth at edgewise apparitions, bright stellar-like points of light are occasionally detectable along the edge of the ring. If present, satellites will normally look like beads of light along the linear

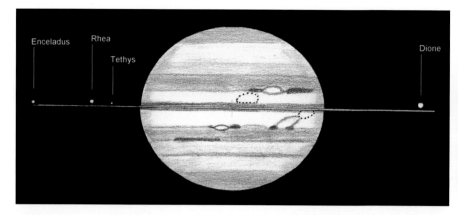

Figure 4.13. Two of Saturn's icy moons are visible as tiny star-like objects in or near the ring plane (from left to right they are Tethys and Dione). North is at the top in this image. Visual observers sometimes see satellites near the plane of the rings at edgewise apparitions, a beautiful spectacle. (Credit: Amanda S. Bosh (Lowell Observatory), Andrew S. Rivkin (University of Arizona/LPL), the Hubble spece telescope High Speed Photometer Instrument Team (R.C. Bless, PI), and NASA.)

extent of the ring, and because of atmospheric turbulence above the Earth's surface, the star-like points may scintillate or twinkle, a very beautiful spectacle as shown in Figure 4.13 (compare this drawing with an analogous image in Figure 4.14).

Figure 4.14. Drawing submitted to the ALPO Saturn Section by Philip Budine of Binghamton, New York, with a 25.4-cm (10.0-in) Newtonian at 350× in good viewing conditions on October 28, 1966, at 02:20 UT. Left to right in the image (W to E) Saturn's moons Enceladus, Rhea, Tethys, and Dione appear near the ring plane. (Credit: Philip Budine; ALPO Saturn Section.)

Of course, beads of light along the dark ring edge are sometimes present without known satellites contributing to their visibility, caused by sunlight passing through major ring divisions, illuminating adjacent ring constituents. Extraplanar ring particles (suspected over the years in Earth-based observations and confirmed by spacecraft) may also be detected from Earth at edgewise presentations of the rings. Extraplanar particles should resemble a faint "haze" above or below the plane of the rings, but the glare from the globe or the rings themselves undoubtedly complicate matters. Finally, when the ring plane passes through the sun, satellite transits, occultations, shadow transits, and eclipses may be visible within or near the equatorial plane of Saturn.

Shadows and Other Globe and Ring Features

Shadow of the Globe on the Rings (Sh G on R)

This feature is normally visible as a dark grayish-black feature on either side of opposition during a given apparition, it is typically regular in form, and deviation from the actual black appearance is due to scattered light and poor viewing conditions (see Fig. 1.3 in chapter 1).

Shadow of the Rings on the Globe (Sh R on G)

This feature is sometimes visible, shown in Figure 1.3, as a dark grayish-black shadow north or south of the rings where they cross the globe (whether the shadow appears north or south of the rings depends on which hemisphere of Saturn is tilted toward our line of sight). Just like the shadow of the globe on the rings, any variation from the true black shadow condition occurs for the same reasons as noted in the preceding paragraph.

Terby White Spot (TWS)

The TWS is an occasionally perceived brightening of the rings immediately adjacent to the Sh G on R (Fig. 4.15). Although it usually appears brighter than all of the zones on the planet's globe and ring components, it is little more than a false contrast effect and not a real feature of Saturn's rings. There is considerable interest, however, in what correlation there is between brightness of the TWS and the varying tilt of the rings.

Occultations of Stars by Saturn's Globe and Rings

Occultations of stars by the rings present unusual opportunities for detecting from Earth the positions of intensity minima in the ring components. Variations in the

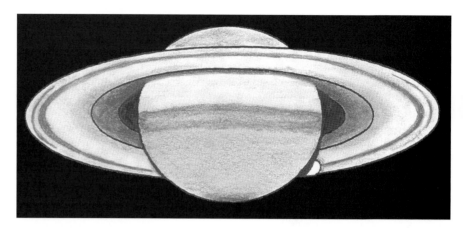

Figure 4.15. To the west (right) of the globe's shadow on the rings is a prominent Terby white spot (TWS) in this drawing contributed to the ALPO Saturn Section by Phil Plante of Braceville, Ohio, using a 20.3-cm (8.0-in) refractor at 333× in excellent viewing conditions on October 22, 1992, at 23:57 UT. The TWS is almost certainly a spurious contrast effect. (Credit: Phil Plante; ALPO Saturn Section.)

brightness of a star during its passage behind the ring system can reveal intensity differences that would not otherwise be perceptible with the same aperture. Unfortunately, only reasonably large instruments are useful in this kind of work, and it is worth noting also that ring A, overall, is much less dense than ring B. For instance, a 15.0-cm (6.0-in) refractor will reveal a star of 7th magnitude (m_v = 7.0) in occultation behind ring A, while a 25.0-cm (10.0-in) instrument will be challenged by an 8th magnitude star (m_v = 8.0). This should not discourage observers with smaller apertures from attempting such observations, but work with any small instrument is necessarily limited to brighter stars that Saturn, unfortunately, seldom occults. Even so, observations that record the times of contact with the planet's limb or ring edges are of great value. Although the star is frequently invisible most of the time the rings are in front of it, the star may suddenly appear when behind an area of lesser density and be visible in the smaller telescope. For a valuable and dynamic record of stellar occultations by Saturn and its rings, high-resolution digital images and "time-stamped" videotape of the entire sequence of events is extremely worthwhile to supplement visual work (Fig. 4.16).

Occultations of Saturn by the Moon

The moon occasionally passes in front of the planet Saturn, and although lunar occultations of solar system objects are of limited scientific interest, they are often spectacular events to watch. Because of the sizeable dimensions of planetary disks, the duration of disappearance and reappearance of Saturn's globe and rings at the lunar limb takes considerably longer than pinpoint objects like stars. Predictions of lunar occultations of Saturn appear regularly in popular magazines and on the Internet, allowing observers to make long-range plans to time the events of disappearance and reappearance. Observers should attempt to record the following

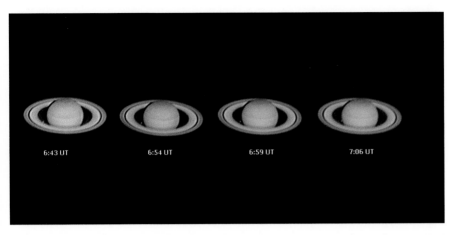

Figure 4.16. On the night of November 15, 2003, between 06:43 and 07:06 UT Rolando Chavez captured these fine images of Saturn and its ring system from Powder Springs, Georgia, as it passed in front of an 8.4 magnitude star in Gemini using a 20.3-cm (8.0-in) Newtonian and a Philips ToUcam. The star is clearly visible in the images as it passed behind ring C (the apparent track of the star is from left to right as Saturn moves W to E in this image). (Credit: Rolando Chavez; ALPO Saturn Section.)

phenomena to the nearest second (all times in UT) whenever possible using a watch synchronized with WWV time signals or an atomic clock:

- Immersion stages:
  1. The last second when the lunar limb makes initial contact with the edge of the rings (i.e., they begin their disappearance behind the moon)
  2. The last second when the lunar limb makes first contact with the edge of the planet's globe (the globe begins its disappearance behind the moon)
  3. The last second when the lunar limb makes last contact with the edge of the planet's globe, which is now behind by the moon
  4. The last second when the lunar limb makes final contact with the edge of Saturn's ring system (the rings and globe are behind by the moon)

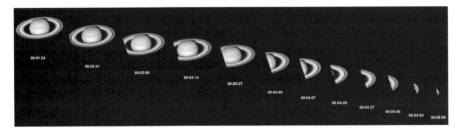

Figure 4.17. On the night of February 20, 2002, Don Parker made this excellent series of CCD images of the disappearance of Saturn behind the dark limb of the moon using a 40.6-cm (16.0-in) Newtonian between 00:01:24 and 00:05:00 UT in good viewing conditions from Coral Gables, Florida. The sequential times of disappearance (immersion) are shown from left to right in the figure. (Credit: Donald C. Parker; ALPO Saturn Section.)

- Emersion stages:
  1. The second when the edge of the ring system first appears at the opposite lunar limb (the rings starts to reappear from behind the moon)
  2. The second when the edge of the planet's globe initially appears at the opposite lunar limb (the globe begins reappearance from behind the moon)
  3. The last second when the lunar limb makes last contact with the edge of the Saturn's globe (reappearance of the globe is now fully complete)
  4. The last second when the lunar limb makes last contact with the outer edge of Saturn's ring system (the rings and globe are now just visible again)

Opportunities to record all of the events of a lunar occultation of Saturn are rare, indeed, and most of the time observers can only capture the initial or final phases of such events because of their geographic location. Sequential high-resolution digital images or videotapes of some or all of an occultation of Saturn, in step with accurate time signals, is perhaps the best way to produce a permanent record of such strikingly beautiful phenomena (Fig. 4.17).

Drawing Saturn's Globe and Rings

Purposes and Objectives of Drawing Saturn

There is no better way to train the eye to detect features, especially the more elusive ones, in the atmosphere of Saturn, or phenomena in the rings, than to make full-disk drawings at the eyepiece. Drawings of Saturn, with accompanying descriptive reports, have a threefold purpose:

1. To keep observers constantly aware of phenomena seen or suspected on the planet's globe and in the rings
2. To establish a reliable, permanent lasting record of visual observations
3. To help develop and maintain an observer's sensitivity to perceive subtle detail at the threshold of vision

Accurate drawings are extremely useful as long-term observations to investigate routine seasonal activity, periodic outbursts (e.g., unusually prominent white spots), and other phenomena that are suspected or readily apparent on Saturn's globe, as well as the relative visibility of various ring components and associated discrete features. Good sketches of the planet serve as permanent documentation of transient bright and dark spots on the globe as a check on rotation periods in different latitudes, presumably in expectation that a complete, reliable ephemeris for System I and II will eventually appear in the *Astronomical Almanac* analogous to that for Jupiter. As discussed earlier in this book, the equatorial regions of Saturn's globe (NEB, SEB, and EZ) have a sidereal rotation period of $10^h14^m00^s$, designated as system I, while the remainder of the globe has a rate of $10^h38^m25^s$, denoted as system II. For polar regions, mainly the SPR and NPR, the rotation rate is normally equal to system I, while system III is a radio rate of $10^h39^m22^s$ corresponding to the interior of Saturn. Observers who use and are familiar with the *Astronomical Almanac* know that system III data for Saturn already are included, but this longitude system applies only to the origin of radio emissions, and there seems little hope that the three systems will turn up in that publication anytime soon. Therefore, organizations like the Association of Lunar and Planetary Observers (ALPO) and British Astronomical Association (BAA) annually publish system I, II, and III data on their Web sites for the convenience of observers. Without question, a great deal remains to be done to confirm rotation rates in different Saturnian latitudes, particularly with respect to monitoring disturbances or spots that persist long enough for central meridian (CM) transits at higher northern or southern latitudes on Saturn. Therefore, the hope is that persistent efforts by Saturn observers will ultimately persuade the Nautical Almanac Office to change its mind and add ephemerides for system I and II longitudes to the *Astronomical Almanac.*

Confirmation of infrequently observed features and suspected detail is a major objective of visual drawings of Saturn. To positively confirm globe and ring features, considerable work by many observers is required; that is, unless several people verify the presence of a spot or festoon, for example, on a given night of observation, it is not easy to claim it as definitely real. To remove subjectivity and provide confirmation of results, an aggressive systematic, simultaneous observing program, where individuals work independently but at the same time and date throughout a given apparition, is of vital importance. More and more individuals who participate in observing programs coordinated by the ALPO and BAA Saturn sections are striving to accomplish simultaneous work, and this collective monitoring of variable phenomena in the atmosphere of the planet and in the rings is accordingly improving the objectivity and reliability of data (an example of near simultaneous observations appears in Fig. 5.1). As established by these and other international team efforts, a more complete, coherent, and realistic picture of Saturn is emerging. As an example, amateur observers have already established from visual observations that faint belts are not just occasionally visible on the

Figure 5.1. Shown here are two pairs of near-simultaneous observations. The first two at left were made less than an hour apart on December 26, 2003, the top CCD image (a) at 22:58 UT by Cristian Fattinnanzi of Macerata, Italy, using a 23.5-cm (9.25-in) Newtonian, and the lower drawing (b) at 23:55 UT by Ivano Dal Prete of Verona, Italy, at 360× using a 20.3-cm (8.0-in) Newtonian. The second pair at right were made on March 1, 2004, the top CCD image (c) at 18:42 UT by Damian Peach of Norfolk, United Kingdom, using a 28.0-cm (11.0-in) SCT, and CCD image (d) at 18:44 UT by Martin Mobberley with a 30.5-cm (12.0-in) SCT observing from Suffolk, United Kingdom. Compare each pair of observations and notice the similarities in appearance of the globe and rings documented by different observers working independently at different geographic locations. (Credits: Cristian Fattinnanzi, Ivano Dal Prete, Damian Peach, and Martin Mobberly; ALPO Saturn Section.)

planet, and we know that faint, delicate features are more frequently perceptible within these belts and zones, as well as in the major ring components, if the proper observational methods are consistently used. It has also been demonstrated that Cassini's and Encke's divisions are not the only such features seen in the rings of Saturn, as confirmed visual sightings of intensity minima in the ring components occur year after year. Furthermore, visual observations have shown that ring C can be seen not only at the ansae but in front of the planet's globe as the dusky Crape ring.

Unfortunately, accurately drawing Saturn and its ring system is considerably more difficult than the other planets because of the rather obvious oblateness of the globe and constantly changing inclination of the plane of the rings from one apparition to another. The ALPO Saturn section furnishes a series of blanks for use during an observing season for drawing the planet correctly with the proper global oblateness and ring tilt. For convenience, a full set of these drawing blanks accompanies this book (see Appendix A), and observers may photocopy the forms for use at the telescope for recording and submitting observations to Saturn sections of organizations such as the ALPO and the BAA, or they are available for download from the ALPO Web site mentioned in the introduction of this book. Each form includes diagrams depicting the proper outline of the globe of Saturn and its accompanying ring system. When selecting the right form to use, careful attention should be given to the value of B, which is the planetocentric (Saturnicentric) latitude of the Earth referred to the plane of the rings (values for B on any given date can be found in an ephemeris, such as the *Astronomical Almanac*). The numerical value of B is positive (+), when northern portions of the globe and ring system are visible from our vantage point, and it is negative (−) when southern areas of Saturn and its rings are seen. B can vary from $0°$ to nearly $\pm 27°$, and when $B = 0.0°$, the rings are exactly edgewise to our line of sight. At $\pm 26.7°$ the rings are open to their greatest extent (tilted toward the Earth at their maximum angle), and the best view of the northern or southern hemisphere of the globe and corresponding face of the rings are possible, as are the extreme polar regions. After entering the value for B onto the selected correct observing form, the next step is to proceed with the drawing itself.

Executing the Drawing

There are a number of fundamental guidelines to follow when making drawings of Saturn and its rings. Prior to coming to the telescope and beginning the sketch, it is always wise to give serious consideration to some basic items that are necessary while observing to avoid last minute scrambling for items. Hare are few of the more essential ones:

- A complete set of pencils with varying sharpness and hardness
- An artist's stump
- Clean erasers and/or erasing pencils of variable sharpness
- A red flashlight (a hiker's headlamp is a great convenience)
- Accurate watch set to WWV time signals heard on shortwave radios at 2.5, 5.0, 10.0, 15.0, and 20.0 MHz or through synchronization with atomic time via the Internet

Once at the eyepiece, and after allowing the instrument's optical system to adjust to the ambient temperature, it is wise to spend a few moments looking at Saturn on a very general basis. It will immediately become apparent if the atmospheric viewing conditions and transparency will allow perception of enough detail to make a drawing feasible. Sometimes it is worthwhile to wait several hours for atmospheric conditions to improve, but if they do not, a detailed written report will often suffice. Pay close attention to the general nature of the globe and rings, and make sure references to any features detected, especially those near the threshold of vision, are clear as to conspicuousness and location within a belt, zone, or ring component. Even though a drawing should represent what an observer realistically sees in the eyepiece, always make a habit of supplementing it with written notes for clarification of anything seen and represented on the sketch. This recurring procedure makes it much easier for someone else to interpret what the observer was trying to show on the drawing.

The beginning phase of making any drawing of Saturn involves establishing the correct relative locations of belts and zones in accordance with the proper latitude, and paying close attention to the overall geometric appearance and actual width of such features. In practice, it is a good idea to sketch in lightly all of the readily apparent belts and zones, and take enough time to get the relative locations and dimensions of these features exact, since rotation does not appreciably distort latitudes. Follow the same procedure for the rings and their components and associated phenomena.

Utilization of the same eyepiece and magnification throughout the drawing session is highly encouraged, but sometimes applying variable magnifications helps to correctly represent features or confirm their presence when they may be questionable at lower powers. Using color filters of known transmission is also extremely beneficial, and always accompany the drawing with careful notes of all equipment and accessories employed and when.

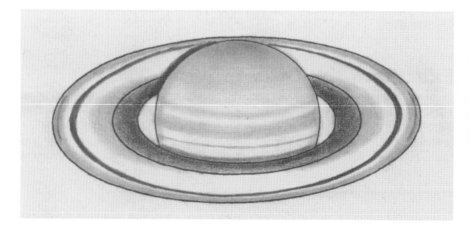

Figure 5.2. This drawing, made by Mike Karakas of Hamilton, Ontario, on October 9, 2002, from 10:30 to 11:00 UT with a 20.3-cm (8.0-in) Newtonian at 240× in excellent viewing conditions. He utilized the basic drawing template provided by the ALPO Saturn Section that represents the correct oblateness of the globe and orientation of the rings on the date of observation. (Credit: Mike Karakas; ALPO Saturn Section.)

Figure 5.3. Some observers make "photographic" drawings, where Saturn appears against a black background, like this one by Mario Frassati observing from Crescentino, Italy, on January 14, 2003, at 21:00 UT using a 20.3-cm (8.0-in) Schmidt–Cassegrain in good viewing conditions. (Credit: Mario Frassati; ALPO Saturn Section.)

Once the fundamental part of the drawing is complete, it is time to enter on the forms the Universal Time (UT) of the "start" of the sketch. With some haste, but without sacrificing accuracy, start filling in the fine details of the drawing. Use dashed lines to set off bright areas or brilliant spots from their immediate environment. Remember to give all areas represented on the drawing equal emphasis, depicting actual and relative appearances. Shading-erasure techniques, often employed in making sketches of lunar features, facilitate finesse and accurate representation of relative intensities and tones for belts, zones, ring components, and integral fine features. About 20 minutes should be all that elapses during this stage of the drawing, because rotation will begin to distort features with increased time. When all of these steps are complete, make a final examination of the drawing to ensure that everything is correct with respect to position, size, and emphasis, and then record the UT of the end of the sketching period (Figs. 5.2 and 5.3).

Strip sketches, not usually warranted for Saturn, may be very useful if features persist long enough for CM transit timings within a particular belt or zone. They are especially helpful for depicting belt or zone features through a span of rotational time. Sectional sketches (Fig. 5.4) are especially valuable as a means of giving particular emphasis to localized phenomena (e.g., spots, disturbances, festoons, unusual regional coloration, etc.) examined under higher magnification, and there is room on the standard CM transit forms for small sectional sketches. Strip and sectional sketches should always accompany the main Saturn drawing form as an attachment.

For every drawing, observers should provide the following supporting data, as illustrated in the example in Figure 5.5:

1. Enter the observer's name, address, and location of the observing station.
2. Include the observer's latitude, longitude, and height above mean sea level.
3. Specity the telescope used, including magnifications, all filters, and accessories.

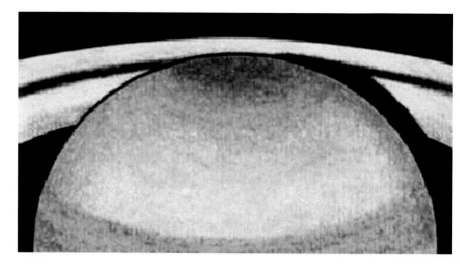

Figure 5.4. Sectional sketches isolate regional phenomena as exemplified by this colorful drawing by Mario Frassati on March 10, 2003, at 20:45 UT using a 20.3-cm (8.0-in) Schmidt–Cassegrain in superb viewing conditions. The intent here was to document a possible overall greenish hue of the SPR and nearby global areas. (Credit: Mario Frassati; ALPO Saturn Section.)

4. Record the field orientation of the oculars (ideally should be the normal inverted and reversed astronomical view, with directions that correspond to the IAU convention).
5. Add values for seeing, D', and transparency, T_r (the quantitative method of evaluation is preferred).
6. Enter the start and end times of the drawing in UT.
7. Enter CM longitudes for systems I, II, and III from an appropriate ephemeris.
8. Make a self-evaluation of the accuracy and reliability of the drawing.
9. Enter the numerical value of B from an ephemeris.

There is room on the standardized drawing blanks for most of this supporting information, and make it a point to include a sufficient, but concise, objective narrative with each sketch. Here are some points to consider in providing such supplementary information:

1. Describe the location, characteristics, and general nature of disturbances, spots, and other phenomena in connection with belts, zones, or ring components.
2. Compare regions of similar nature and latitude in opposite hemispheres (whenever possible); for example, compare the SEBs with the NEBn. It is also possible to compare areas that are adjacent or similar in the same hemisphere (e.g., SEBs vs. SEBn).
3. Make general notes regarding the prominence of belts, zones, or ring components.
4. Note the general width and extent of the ring shadow on the globe and the shadow of the globe on the rings.
5. Describe the visibility and location of any "intensity minima" in the rings, in addition to studies of Encke's, Cassini's, and possibly Keeler's divisions.

Association of Lunar and Planetary Observers (A.L.P.O.): The Saturn Section
A.L.P.O. Visual Observation of Saturn for B = –26° to –28°

S

N

Coordinates (check one): [ ] IAU [√] Sky

Observer___*Phil Plante*___ Location___*Poland OH*___

UT Date (start)___*De 27 2003*___ UT Start___*2ʰ 20ᵐ*___ CM I (start)___*317.2*___° CM II (start)___*339.9*___° CM III (start)___*318.4*___°

UT Date (end)___*Dec 27 2003*___ UT End___*2ʰ 38ᵐ*___ CM I (end)___*327.8*___° CM II (end)___*350.0*___° CM III (end)___*328.5*___°

B=___*⁻25.4*___° B'=___*25.6*___° Instrument___*8" SCT (6-8)*___ Magnification(s)___*333*___ Xmin _____ Xmin

Filter(s) IL(none)_____ f₁___*38A*___ f₂___*23A*___ f₃_____ Seeing___*5, 7momeul*___ Transparency___*+4.0 (SI. HAZe)*___

Saturn Global and Ring Features	Visual Photometry and Colorimetry				Absolute Color Estimates	Latitude Estimates ratio y/r
	IL	f₁	f₂	f₃		
Ring A10-5	7.0	7.5	7.0			
A 5-0	7.5	7.5	7.0			
B10-7	8.0	8.0	8.0	—— Rep.		
B 7-0	7.6	8.0	8.0			
C10-3	0.7	0.3	0.1			
C 3-0	0.0	0.0	0.0			
EZ	7.6	6.5	8.0			
SEBn	6.0	6.0	6.5			
SEBs	6.8	6.0	6.5		SEBn N edge	-0.3
STrZ	7.8	6.0	7.0		SEBs S edge	+0.0
STeB	7.5	6.0	6.0		STeB	+0.5
STeZ	7.8	7.0	7.0		SPR N ed.	+0.65
SPR	6.5	5.8	6.0		SPC N edge	+0.85
SPC	5.0	5.8	5.8			
CRepe Band	1.0	—	0.5	—— access Glabe @ B0-C8		
				Sketch made in IL		

<u>Bicolored Aspect of the Rings:</u>   No Filter   (IL   ) (check one):   [√] E ansa = W ansa [   ] E ansa > W ansa [   ] W ansa > E ansa
(always use IAU directions)   Blue Filter (_38A_) (check one):   [√] E ansa = W ansa [   ] E ansa > W ansa [   ] W ansa > E ansa
                                Red Filter   (_23A_) (check one):   [√] E ansa = W ansa [   ] E ansa > W ansa [   ] W ansa > E ansa

IMPORTANT: Attach to this form all descriptions of morphology of atmospheric detail, as well as other supporting information. Please <u>do not</u> write on the back of this sheet. The intensity scale employed is the *Standard A.L.P.O. Intensity Scale*, where 0.0 = completely black ⇔ 10.0 = very brightest features, and intermediate values are assigned along the scale to account for observed intensity of features.   Copyright © 1996 Form S–2628 JLB

Figure 5.5. Phil Plante of Braceville, Ohio, made this sketch using an ALPO Saturn Section observing form on December 27, 2003, between 02:20 and 02:38 UT with a 20.3-cm (8.0-in) Schmidt–Cassegrain at 333×. All of the pertinent data appear on the form, including intensity and latitude estimates (discussed later in this book). (Credit: Phil Plante; ALPO Saturn Section.)

6. Include notes on the bicolored aspect of the E and W ring ansae, paying attention to whether or not similar effects are visible on the corresponding global limbs, and take note of any azimuthal brightness asymmetries in the rings.

7. Note the general appearance and nature of the rings at edgewise presentations (when $B = 0°$).

8. If drawing Saturn in integrated light (no filter) and with color filters, be sure to identify in the notes all filters used and what features appear on the drawing that are enhanced in different wavelengths.

9. Report anything unusual or of particular significance, especially if not shown on the drawing or sectional sketch.

10. Make reference to and include images that were captured on the same date that the drawing was completed (comparing images to drawings made on the same night is always meaningful).

Nomenclature and Field Orientation of the Image

As discussed briefly in Chapter 1, the nomenclature system for Saturn's rings and globe (Fig. 1.3) as well as the relevant abbreviations and symbolism used (see Table 1.1), are important terms to commit to memory. In Figure 1.3 south (S) is at the top and west (W) is to the left in the sky by using a telescope that inverts and reverses images (the normal astronomical telescopic view), but written descriptions of Saturn (and directional indicators on drawings) should adhere to the IAU system (see Chapter 2). This means that east (E) is toward the left (true E on the planet), as indicated in Figure 1.3, in the normal inverted astronomical view.

Recall from discussions earlier in this book that ring A is the usually seen outermost component of Saturn's rings, ring B occupies the middle portion of the ring system, and ring C is the innermost visible component. Within ring A are the gaps of Encke and Keeler, while situated between rings A and B is Cassini's division. To provide a means for accurate and easy identification of ring divisions and more subtle intensity minima, a convenient system exists for denotation of such features. Assign a capital letter and a number to the division or intensity minimum seen. The letter denotes what ring component the feature is located in, and the number indicates the relative position of the feature as a fraction of the distance outward from the globe into the ring. Encke's division, for example, is designated E5 because it is about halfway out from the globe of Saturn in ring A, while Keeler's gap is denoted as A8 because it is about 80% of the way out in ring A. Likewise, Cassini's division is A0 or B10 because it is at exactly at the boundary between rings A and B.

When looking at Saturn in a telescope that inverts and reverses the image (as seen in Chapter 1, Fig. 1.3), the globe casts a shadow on the ring system to the left or IAU east prior to opposition, to the right or IAU west after opposition, and on neither side exactly at opposition (no shadow). The series of images in Figure 5.6

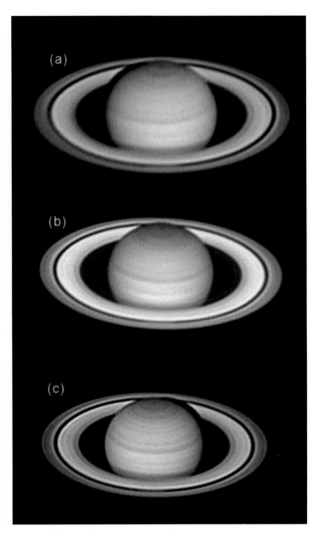

Figure 5.6. The shadow of the globe on the rings (Sh G on R) shifts from E to W of the globe as Saturn progresses through opposition. Look at the following images made by Damian Peach Norfolk, United Kingdom, using a 23.5-cm (9.25-in) Schmidt–Cassegrain and a Philips ToUcam: (a) October 16, 2002, 02:48 UT (prior to opposition); (b) December 18, 2002, 23:43 UT (near opposition), and (c) February 19, 2003, 19:48 UT (after opposition). (Credit: Damian Peach; ALPO Saturn Section.)

illustrates this important point, and observers should always try to correctly represent shadows on drawings.

It is worthwhile to reiterate here that field orientation is probably one of the most important things to get right when drawing Saturn. In a standard astronomical telescope (no prismatic star diagonal or other right-angle device to confuse image orientation), the sky direction of south is at the top of the field of view while west is to the left. This corresponds to a field lying on the celestial meridian between the zenith and the south point of the horizon in the Northern Hemisphere of the Earth. After reducing observations at the end of many apparitions, I have seen a considerable number of instances where errors in field orientation occurred in descriptive notes and on drawings due the use of star diagonals. Try to avoid using any device that reorients the telescopic image, even though this practice may hamper viewing comfort to some degree.

Bear in mind that sky directions are not necessarily those of the International Astronomical Union (IAU) convention on all moons and planets; therefore, the IAU General Assembly in 1961 adopted a resolution whereby directions in astronomical literature must correspond to true directions on the planet observed. For consistency, observers should always use the IAU system of reference as shown in Chapter 1, Figure 1.3, when making and reporting Saturn observations; so that there is absolutely no confusion, at all times draw Saturn as it appears in the normally inverted and reversed view using the IAU convention (south at the top and east to the left). Because the drawing blanks in Appendix A are suitable for sketching Saturn with values of B ranging from $0°$ to $\pm26.7°$, it is necessary to invert the forms to draw Saturn correctly (as seen in the telescope) with south at the top when the northern hemisphere of the globe and corresponding face of the rings are inclined toward Earth. The blanks also have a space for the observer to indicate whether IAU directions or sky directions are applicable. Furthermore, in planetary science, prograde motion is from west to east, and features move from right to left across Saturn's globe in the normal inverted and reversed telescopic view. Note that the following, f, limb of Saturn is west in IAU terminology; preceding, p, is east in the IAU sense.

Factors Affecting Drawing Reliability

Observers should be aware of several things that can ultimately affect their drawing accuracy. One of these pitfalls, known as repetitive style, is easy to avoid through experimentation with drawing form and technique. Another is depicting excessive sharpness for boundaries between belts and zones, none of which can really be resolved visually with edges that are precisely linear and crisply defined. To minimize this problem when sketching Saturn's belts and zones, observers should use a relatively blunt pencil when drawing. Differences in transparency and seeing affect drawings, too, and the best solution to this dilemma is to choose viewing times when the seeing and sky transparency is above average, as well as observing when Saturn is as high above the horizon as possible.

Visual acuity, whereby one person simply sees better than another, is a factor that is hard to measure and correct for. The aperture of the telescope is critical, so the proper choice of instrument size generally dictates how much detail can be resolved. For example, observations with too small an aperture usually result in a drawing that is markedly simplified. Larger instruments usually resolve more detail to sketch, all things being equal, and comparing drawings made with small apertures to those done with larger instruments can be very troublesome for an analyst trying to reduce many different observations at the end of an apparition. Always exercise care when appraising atmospheric conditions, and employ the best aperture and magnifications that produce good image brightness and contrast. Include in the observing record notation of any anomalies with regard to personal visual acuity. One way to help overcome the latter problem is to evaluate the drawing as objectively as possible; that is, the level of confidence placed in the final drawing should reflect limiting visual deficiencies. Astigmatism, for example, is the worst of the visual defects of the eye, and observers suffering from this affliction should wear their corrective lenses when observing and drawing Saturn.

Contrast sensitivity is frequently variable to some degree from one person to another. Those who suffer from especially poor contrast sensitivity often make drawings that consistently lack subtle tonal differences and are devoid of intricate detail when compared with sketches by others having normal contrast perception using the same aperture. The faintest features on Saturn are apparently beyond the visual threshold of such individuals with a given instrument, and a slight increase in aperture and magnification may help those plagued by an inability to see atmospheric or ring detail. Observers with exceptionally good contrast sensitivity and perception, on the other hand, often exaggerate tonal differences, and their drawings are often misleading. Experience has shown that an observer who is bothered by what appears to be an apparent inability to represent properly with a pencil a great range of tonal differences probably has good or near-optimum contrast perception. Indeed, as one gains experience in observing and drawing Saturn, contrast sensitivity shows marked improvement, so novice observers should be patient.

Color sensitivity is yet another factor that affects drawings of Saturn. It is clear that some eyes are particularly sensitive to specific wavelengths and less responsive to others. An observer with a strong sensitivity toward blue wavelengths, for example, may believe that a bluish zone on Saturn is the brightest zone on the globe, while another individual may have a greater affinity for yellow light and feel that a yellowish zone is more brilliant. If an observer is aware that he has psychophysical color sensitivity at a certain wavelength, he should not forget to make a note of it in his descriptive report. Color-blind individuals are particularly handicapped, and they should not attempt absolute color estimates.

Those who make errors systematically with regard to proportion may draw features too small or too large relative to one another or to the planetary globe or rings. If the magnification is too low, irradiation might produce an effect whereby dark areas appear smaller in proportion to lighter regions, and magnifications that are too high sometimes cause an observer to draw dark features or regions too large in relation to the disk of the planet. Positional errors significantly affect drawings too, so exercise care in establishing the location of a feature relative to another one. It is easy to correct or reduce systematic errors in proportion and position by making simulated sketches of the planets and comparing the results before coming to the telescope.

Fatigue and distractions are factors that often critically affect drawings and data acquisition. Observers should always get plenty of rest before arriving at the telescope, and many experienced observers "pre-sleep" for a period of time that is equal to the anticipated observing run. Try also to promote good concentration and attention to detail during the observing session by removing distractions such as cellular telephones and CD players.

Even though observing equipment and methods are rapidly evolving and becoming more sophisticated, visual systematic (and simultaneous) efforts in drawing and describing Saturn will always have enduring importance. Nevertheless, as in any field of observational work, it is worthwhile to continually examine, criticize, refine, and develop drawing methods and techniques to as high a level of precision as possible. There is no question that amateur visual observations in the form of well-executed drawings have long been a major part of our body of knowledge about Saturn and its ring system. Despite the recent emergence of CCD and webcam imaging, comparative quality sketches of Saturn by skilled observers will continue to be a vital subset of any truly comprehensive observing program. Indeed, there is no greater or more unforgettable experience than seeing Saturn and its majestic rings "in the raw" with the human eye through a quality telescope in superb viewing conditions.

Methods of Visual Photometry and Colorimetry

Visual Numerical Relative Intensity Estimates (Visual Photometry)

A thorough and continuous record of variations in the relative intensities of the different belts, zones, and ring components of Saturn is a valuable data source on any global seasonal atmospheric phenomenon and other fluctuations over time. Observers should make it standard procedure to carry out regular intensity estimates while at the telescope, since there is a definite and rather uniform relationship between relative intensities and the real albedo values of Saturn's features. Furthermore, after capturing a series of images of Saturn with a CCD camera or webcam during a given observing session, it is crucial not to forget to perform visual estimates of belt, zone, and ring component intensities on the same night.

The Association of Lunar and Planetary Observers (ALPO) Saturn Section data gathered over the last 30 years (which roughly corresponds to one Saturnian year that spans 29.5^Y on Earth) demonstrate that relative intensities of different belts and zones on the planet's globe seldom remain constant from one apparition to the next. Because of its obliquity of $26.7°$ (the angle between its axis of rotation and the pole of its orbit), one global hemisphere is tipped toward or away from the sun as it moves along its orbit. Thus, Saturn may present the best opportunity in the solar system for determination of subtle seasonal effects on belts and zones, since Jupiter has virtually no seasons at all. Because the seasons on Saturn are so long by terrestrial standards, a project like this is unavoidably a very lengthy pursuit, but again, there is considerable observational evidence pointing to a delicate seasonal effect.

For the sake of consistency in making systematic visual numerical relative intensity estimates (visual photometry) at the telescope, it is important for all observers to employ a suitable reference standard, the best example of which is the ALPO Relative Numerical Intensity Scale. This scale consists of a numerical sequence from 0.0 (totally black or shadow condition) to 10.0 (most brilliant white condition), and intermediate values are assigned to features on Saturn to the nearest 0.1, whenever possible, using the scale. Specifically for observations of Saturn, it has been necessary to modify this scale as the ALPO Saturn Relative Numerical Intensity Scale. It is identical to the aforementioned scale except that the outer

Table 6.1. Standard saturn visual numerical relative intensity scale

Numerical value	Descriptive interpretation	Representative Saturnian features
10.0	Brilliant white	Brightest features of all
9.0	Extremely bright	Very brightest objects
8.0	Very bright	Very bright zone or ring—outer third of ring B (standard)
7.0	Bright	Ordinary bright zone or ring
6.0	Slightly shaded	Dull zone
5.0	Dull	Dull zone; typical polar regions
4.0	Dusky	Polar regions; dusky belt
3.0	Dark	Ordinary dark belt
2.0	Very dark	Very dark belt
1.0	Extremely dark	Unusually dark features
0.0	Completely black	Shadows

third of ring B is the adopted standard on the numerical sequence. The outer third is the brightest part of ring B, and it has a stable intensity of 8.0 in integrated light (no filter); thus, intensity estimates of all other features on the globe and in the rings are made using this standard of reference for most apparitions when ring B is clearly visible. With practice, a very consistent, fairly objective, and accurate series of intensity estimates are usually achievable, but the real key to success and long-term data reliability is regular usage of the scale when observing Saturn. Table 6.1 gives the numerical intensity values established for Saturn and its ring system, along with some comparative references for intensities of certain specific belts and zones.

It is always wise to avoid lengthy verbal descriptions of intensities, so make it a practice to rely on the numerical value assigned to features as an indication of their relative brightness. There is room on the standard observing forms for entering estimated visual numerical intensity values (see Appendix A).

The process of making reliable visual numerical relative intensity estimates is quite simple in actual practice. At the eyepiece, in integrated light (no filter), simply list all of the features seen (e.g., belts, zones, ring components, shadows, etc.) in order of decreasing brightness in a notebook. Compare the brightest feature with the standard of reference (the outer third of ring B) and assign to it a relative numerical intensity. Follow the same procedure for the rest of the features on the list (brightest to darkest), taking care to make all estimates and comparisons as precisely as possible. Once the list is complete, which should show global features in order of decreasing brightness, transfer the tabulation of estimated values to the standard Saturn observing form (Fig. 6.1). Follow the same process for estimates of the ring components and enter them on the form as well, and record localized spots and features in the same manner as the principal belts, zones, and ring components. Several belts near the equatorial regions of Saturn show multiplicity on occasion (e.g., SEBn and SEBs), so assign these belt components individual intensities. Identify and estimate divisions or intensity minima in the rings using the same technique.

So that clear interpretations of relative numerical intensity estimates can occur, always make an effort to optimize contrast sensitivity and perception using methods described earlier in this book. Strive to minimize any psychophysical effects associated with small image size and low surface brightness by using proper apertures and magnifications, all under the best viewing conditions possible.

Association of Lunar and Planetary Observers (A.L.P.O.): The Saturn Section
A.L.P.O. Visual Observation of Saturn for B = –26° to –28°

S

P 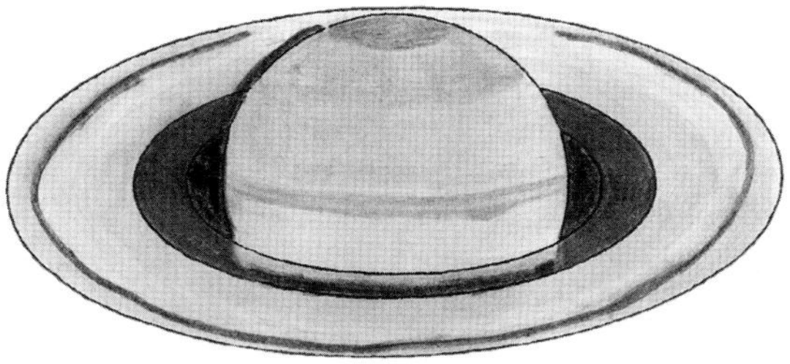   F

N

Coordinates (check one): [X] IAU [   ] Sky

Observer _____Carl Roussell_____ Location _____Hamilton Out Can_____ 43° 15' N   79° 49' 55" W

UT Date (start) _Dec 1, 03_ UT Start __9:00__ CM I (start) _195.7_ ° CM II (start) _330.2_ ° CM III (start) _339.2_ °

UT Date (end) _Dec 1, 03_ UT End __9:50__ CM I (end) _225.0_ ° CM II (end) _358.4_ ° CM III (end) _7.4_ °

B= _25.1_ ° B'= _25.8_ ° Instrument _____15cm f/8 RL_____ Magnification(s) _123_ Xmin _204_ Xmin

Filter(s) IL(none) ___√___ f₁ __12__ f₂ __58__ f₃ __21__ Seeing _2-3_ Transparency __3__

Saturn Global and Ring Features	IL	f₁	f₂	f₃	Absolute Color Estimates	Latitude Estimates ratio y/r
	Visual Photometry and Colorimetry					
Ring A	7	7	6	6	Yellow white	
Ring B (outer ⅓)	8	8	7	7	White	
Ring B (inner ⅔)	7	7	6	6	Yellow white	
Ring C	1	1	0	1	dark gray	
G Shon R	0	0	1	0	black	
R Shon G	0	0	0	0	black	
Cassini	0	0	0	0	black	
EZ	6	7	5	7	Yellow white	
SEB tot	4	5	4	4	brown	
SEB n	4	4	4	4	brown	
SEB z	5	6	5	5	Yellow brown	
SEB s	4	5	4	4	brown	
STc region	4	4	3	2	blue gray	
SPC	3	3	2	1	greenish gray	
Remainder Globe	5	6	4	5	yellow	
Globe s of SEB	/	/	/	4	blue gray	

Bicolored Aspect of the Rings: No Filter (IL) (check one):   [√] E ansa = W ansa [   ] E ansa > W ansa [   ] W ansa > E ansa
(always use IAU directions)   Blue Filter _80A_ (check one):   [√] E ansa = W ansa [   ] E ansa > W ansa [   ] W ansa > E ansa
   Red Filter _25_ (check one):   [√] E ansa = W ansa [   ] E ansa > W ansa [   ] W ansa > E ansa

IMPORTANT: Attach to this form all descriptions of morphology of atmospheric detail, as well as other supporting information. Please do not write on the back of this sheet. The intensity scale employed is the *Standard A.L.P.O. Intensity Scale*, where 0.0 = completely black ⇔ 10.0 = very brightest features, and intermediate values are assigned along the scale to account for observed intensity of features.   Copyright © 1996
Form S–2628 JLB

Figure 6.1. Carl Roussell, observing from Hamilton, Ontario, recorded visual numerical relative intensity estimates in integrated light (no filter) as well as with color filters on December 1, 2003, at 09:00 to 09:25 UT using a 15.2-cm (6.0-in) Newtonian at 204×. (Credits: Carl Roussell; ALPO Saturn Section.)

Systematic errors, which are essentially unavoidable, can be corrected by identifying and quantifying personal equations. Observers should try to ascertain what differences exist between their own work and that of others (which might come to light as part of a simultaneous observing program), with the recognition of how high or low personal estimates may be in reference to those made by more experienced individuals using similar equipment and observing under analogous conditions. Conversely, random errors are much harder to detect and alleviate by correction, although the law of averages has a way of reducing these to a minimum.

During apparitions when Saturn's rings are edgewise to our line of sight, ring B is not always visible and available as the intensity standard, forcing observers to resort to an interim point of reference. The key is to select an alternative reference standard that has not shown a pattern of sizeable brightness fluctuations over time. More often than not, the EZ (at an assumed value of 7.0) has traditionally been the logical selection, despite the fact that this feature varies more widely than the outer third of ring B from apparition to apparition (e.g., white spots have appeared in this region from time to time, affecting its intensity). Alerts are issued by the ALPO Saturn Section prior to minimal ring inclinations so that observers are aware of what area is chosen as the short-term intensity standard (as well as what numerical value will be assigned to the feature).

Although the standard ALPO Saturn Relative Numerical Intensity Scale is widely employed throughout the world, some intensity scales differ considerably from this scheme, which can lead to confusion in making a comparative analysis of a large volume of intensity estimates from different sources. For example, the British Astronomical Association (BAA) employs a scale from 0.0 (brightest features) to 10.0 (darkest features like shadows), exactly the reverse of the one used by the ALPO, with no specific intensity reference point assigned for Saturn. While efforts have been underway to normalize scales internationally, there has only been limited success in coming up with a scale that best suits everyone. Whenever possible, all observations contributed by observers who participate in the programs of the ALPO Saturn Section should use the ALPO scale. The most experienced observers who contribute data to both the ALPO and the BAA Saturn sections are often very accommodating and routinely make separate visual numerical relative intensity estimates using the two scales. Data reduction is tedious and rather cumbersome when trying to convert from one scale to another, although computer programs exist that accomplish this task reasonably well.

Filter Techniques (Visual Colorimetry)

Comparison of the reflectivity of various regions of Saturn's atmosphere in different wavelengths of light (visual) constitutes one of the more important methods of studying the planet available to the planetary observer, since the way that reflected light emanates from different atmospheric regions provides valuable clues as to its chemical and physical properties.

Given that a specific color filter will afford transmission in only a very definite range of the visual spectrum, preventing at the same time the passage of other wavelengths, the most useful and readily available means with which to attempt visual colorimetry of Saturn is by employing color filters of precisely determined wavelength transmissions. Such color filters are of tremendous importance to observers because they help differentiate between light reflected from various levels of Saturn's atmosphere, and they also provide a means of improving contrast between regions of dissimilar hue and help minimize image deterioration resulting from atmospheric scattering of light and dispersion.

Wratten color filters, manufactured and distributed by Eastman Kodak, are the most readily available and highly recommended types of filters. They are obtainable in a variety of forms (e.g., optical glass and gelatin film), are inexpensive, have accurate wavelength characteristics, and their color stability persists a long time. Table 2.1 in Chapter 2 lists the more frequently encountered Wratten filters among planetary observers.

As we discussed in Chapter 3, the retina of the eye consists of two basic types of nerve endings that are light sensitive: the rods and the cones. The rods are responsive only to variations in the intensity of illumination, and they are responsible for night vision. Daylight (photopic) vision, therefore, is due to the activity of the rods in relation to differences in light intensity. The cones, however, are specifically receptive to color sensations, and they are responsible for color vision. The rods and cones are active simultaneously under a variety of light conditions, but their functions are clearly not identical.

The normal range for visual sensations of the human eye is from 3900 to 7100 Å, although it is well to remember that this range may vary among people. Maximum visual sensitivity is attained at about 5500 Å (yellow-green light), but as the brightness level diminishes, the optimum sensitivity point shifts toward shorter wavelengths (i.e., toward the blue end of the spectrum). This peculiar effect, as we saw earlier, is the Purkinje phenomenon.

Color sensitivity is influenced by various physical conditions existing within the eye itself, with the color or wavelength of light, and with image brightness. As the aperture of the instrument is increased, there is a corresponding amplification of the apparent image brightness with respect to magnification. Thus, the color response of the cones is improved. Since it appears that color sensations result from composite reactions of the red-, green-, and blue-sensitive cones, it is possible to observe on one of the three cone-sensitive wavelengths by employing a single filter that has a dominant wavelength close to the natural response of the cone. The color of any visible feature on Saturn can be determined by comparing its intensity as viewed separately with red, green, and blue filters. Therefore, it is important to use color filters in making visual relative numerical intensity estimates of planetary features.

Color filters, as mentioned before, are extremely useful in reducing the effects produced by scattered light in planetary atmospheres. Blue wavelengths are scattered more than others, and when an observer tries to look for a feature deep in Saturn's atmosphere, difficulties arise. Because the eye is particularly sensitive to shorter wavelengths when it is dark-adapted, it is useful to filter out the atmospheric violet and blue light with a red or yellow filter when attempting to see surface details on a planet. In the case of Saturn, which has a relatively dense atmosphere, observations with blue filters tend to show features that lie a bit higher in the atmosphere of the planet than could be detected with red or yellow filters.

Color filters are also helpful for reduction of the effects of atmospheric dispersion, so when Saturn is near the horizon, a red or yellow filter will help minimize the spurious color inherent in the image. Irradiation, a contrast effect between areas of significantly different brightness levels, usually impairs resolution when the image is very bright. An increase in magnification will frequently improve the

Table 6.2. Recommended tricolor filters series by aperture

Apertures <15.2 cm (6.0 in):	W23A (red-orange)
	W57 (green)
	W38A (blue)
(The above filters have less density and are more useful with smaller aperture)	
Apertures ≥15.2 cm (6.0 in):	W25 (red)
	W57 (yellow-green)
	W47 (deep blue)
(The above filters are more dense and useful with low powers or larger apertures)	

quality of the image, but it is usually preferable to employ a low transmission filter or perhaps a good variable-density polarizer instead of an indiscriminate use of magnification.

The technique of visual planetary colorimetry involves attempting relative numerical intensity estimates using color filters of known transmission and density. The standard ALPO Saturn Relative Numerical Intensity Scale should always be employed first in integrated light (no filter), with the outer third of ring B having an assigned intensity of 8.0 as before. Follow-up estimates made in integrated light with filter observations using the same magnifications (refer again to the example in Fig. 6.1). Recommended tricolor series of Wratten color filters with reference to specific aperture ranges appear in Table 6.2.

Although Table 6.2 offers explicit recommendations, when planetary surface brightness is relatively low in a larger telescope, it makes sense to use a series of filters that are better suited for smaller apertures. A W23A (red-orange) filter will enhance red and yellow features, while darkening bluish areas. A W38A (blue) filter will deliver a reversal of the effect noted with the W23A (red-orange) filter, darkening reddish or yellowish belts and zones on Saturn. A W57 (yellow-green) filter improves contrast by darkening bluish and reddish belts or zones. A comparative study with the three tricolor filters often produces interesting results. Figure 6.2, illustrates subtle differences in the visibility of global and ring features in different color filters.

A W30 (light magenta) filter, often called the "universal color filter" because it transmits both red and blue wavelengths while suppressing green light, substantially increases the threshold detection of low contrast markings on Saturn. Meaningful results have been obtained on other planets as well because of the remarkable versatility of this filter.

A W82 (light blue) color filter is extremely useful for visual latitude work on Saturn because it increases image contrast and sharpens boundaries between reddish and bluish features on the planet. At the same time, it does not appreciably reduce image brightness nor does it affect to any degree the visibility of the limb of Saturn.

For colorimetric work, use achromatic refractors with caution because of the likely presence of spurious color, especially those of large aperture with short focal lengths (although filters are now available that effectively block many of the detrimental effects of the secondary spectrum). Reflectors and catadioptrics are generally better for colorimetry, but the alignment and optical quality must be good. Larger apertures perform best when viewing conditions cooperate because they produce images of Saturn with greater apparent surface brightness. It is best to

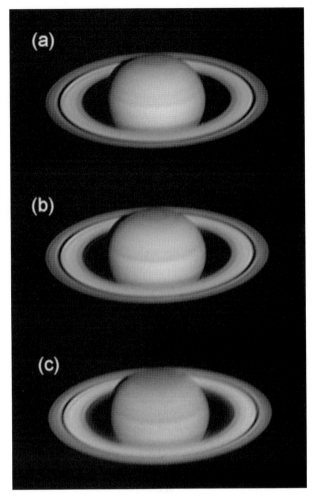

Figure 6.2. This group of images made by Jason P. Hatton of Mill Valley, California, on October 18, 2003, at 12:39 UT using a 23.5-cm (9.25-in) SCT and a Philips ToUcam webcam aptly illustrate the subtle differences on Saturn's globe and rings revealed using (a) red, (b) green, and (c) blue filters. Experienced visual observers are quite familiar with the enhancements that color filters of known wavelength transmission can provide when viewing Saturn. (Credits: Jason P. Hatton; ALPO Saturn Section.)

carry out observations when viewing and transparency conditions are excellent with a fully dark-adapted eye, as well as when Saturn is more than about 30° above the horizon to minimize atmospheric dispersion effects. Shifting vision frequently throughout the available field of view and across the planetary disk helps improve perception of minor differences in hue and contrast. Standard observing forms, such as those available from the ALPO Saturn Section, have provisions for recording colorimetric data (see Appendix A).

Absolute Visual Color Estimates

Absolute visual color estimates are useful if perceived hues on Saturn are compared systematically with a satisfactory color standard in integrated light (no filter), and observers should have normal color vision for this type of work. A color standard used by the ALPO Saturn Section employs colored paper wedges, available from many local art supply stores, for comparative referencing. For best

Table 6.3. Standard color abbreviations for Saturn observations

Color	Abbreviation	Color	Abbreviation	Color	Abbreviation
Brown	Br	White	W	Grayish-black	Gy-Bk
Blue	Bl	Black	Bk	Yellow-white	Y-W
Gray	Gy	Orange	Or	Reddish-brown	R-Br
Red	R	Bluish-gray	Bl-Gy	Reddish-orange	R-Or
Green	Gr	Yellow-orange	Y-Or	Orange-brown	Or-Br
Yellow	Y	Bluish-black	Bl-Bk	Grayish-white	Gy-W

results, illuminate the colored wedges with a small tungsten lamp whose light is filtered through a W78 color filter, but this arrangement may not be practical at the telescope for all observers. Because of the changing availability of reliable standard color wedges, consult the ALPO Saturn Section for the most up-to-date list of suppliers.

Simplicity in making descriptive notes of color is always desirable, and the abbreviations given in Table 6.3 should make it easier to record concise color information on standard observing forms. For nonstandard color abbreviations—those not listed in Table 6.3—it is very important to clarify in observing notes what the notations describing a particular hue mean. Some observers feel it is worthwhile to draw Saturn with colored pencils as a substitute for making absolute color estimates, but despite their artistic value, such drawings are often very misleading and hard to calibrate. Because it is extremely difficult to standardize such work, sketching in color is not usually a recommended practice.

Investigating the Bicolored Aspect of Saturn's Rings

An occasional attention-grabbing, yet poorly understood, phenomenon associated with the rings of Saturn is the bicolored aspect of the rings, where the west and east (IAU) ansae exhibit unequal brightness characteristics when viewed in integrated light, then alternately with red (W23A or W25) and blue (W38A or W47) filters. Observers have recorded this effect when Saturn has been near the zenith where prismatic dispersion effects are minimal, and simultaneous observations have been received that confirm this unequal brightness effect of the ring ansae when there is no corresponding anomaly visible on Saturn's globe.

Saturn observers also periodically report visual suspicions of another strange variance associated with the rings, namely azimuthal brightness and color variations in ring A. Just like the bicolored aspect, they show up from time to time unconnected to equivalent effects on Saturn's globe (which one would expect to occur if atmospheric dispersion was at fault). Such azimuthal asymmetries in brightness and hue apparently arise when light is scattered by denser-than-average particle agglomerations orbiting in ring A, and simultaneous confirmation of these variations in integrated light and with color filters is very worthwhile.

Interestingly, the bicolored ring aspect and azimuthal asymmetries associated with the rings have purportedly been imaged successfully on a few occasions, perhaps captured in the images shown in Chapter 4, Figure 4.10, and in Figure 6.3,

Figure 6.3. Circumferential variations in hue are presumably apparent in ring A in this CCD webcam image taken on January 10, 2003, at 20:49 UT by John Sussenbach of the Netherlands using a 28.0-cm (11.0-in) SCT. Are these differences in brightness and hue real or illusory phenomena? (Credit: John Sussenbach; ALPO Saturn Section.)

and for many, it is tempting to believe that they are real phenomena. Although not fully explained, these very curious spectacles certainly merit further serious study by visual observers of the planet.

Observers who want to image Saturn systematically in search of the bicolored aspect or the strange azimuthal brightness differences in ring A may do so by employing CCD and digital cameras or webcams to perform systematic patrols of the planet in integrated light, and then follow up using red and blue Wratten tricolor filters (see Table 6.2). Since CCD chips are exceptionally sensitive to infrared (IR) wavelengths, which come to focus at a slightly different position than those of visible light, images may often appear fuzzy with poor color saturation, especially with refractors and catadioptrics. The remedy for this problem is to insert an IR blocking filter that threads onto the adapter attaching the webcam or digital camera to the telescope, just like any other eyepiece filter (CCD and digital imaging are discussed in Chapter 9). After downloading the raw images to a computer, it is worthwhile to compare those frames made in integrated light with the ones captured using red and blue filters. Adjusting brightness and contrast while looking at the images on a computer monitor may help reveal subtle differences in hue seen in different wavelengths that may be indicative of the bicolored aspect or azimuthal brightness asymmetries in ring A.

A suggested strategy by the ALPO Saturn section for possibly improving the chances of detection and quantification of the curious bicolored aspect of Saturn's rings, as well as azimuthal brightness asymmetries in ring A, involves monitoring extremely small brightness fluctuations in different wavelengths at various position angles around the ring system. This technique involves putting to use "old-style" photoelectric photometers of 1970s to early 1980s vintage with an aperture of roughly 1.0″ to 2.0″ using standard Johnson UBV filter sets (or perhaps just B and V). Applied to Saturn's rings, this technique admits a considerable amount of light required for very precise magnitude measurements. Observations at 10° to 15° intervals in position angle all the way around ring A, calibrated with ring B measurements, can be repeated at time intervals of 20^m to 40^m on a single night. Such measurements may capture progressive changes in the rings, and perhaps conjectured underlying structures, as they revolve around Saturn. Routine observations of this kind throughout several apparitions may provide quantitative information, currently lacking, which theorists need for model building. Older photoelectric photometers, that some consider outmoded by CCD photometers due to their relative ease of use, can regain their usefulness in observational astronomy in such projects. The ALPO Saturn section is seeking observers who still possess these vintage photoelectric photometers who want to participate in such a long-term project. Moreover, professional and university observatories may own top-of-the-line photoelectric photometers left over from the 1970s or 1980s that may serve a worthwhile purpose.

Determining Latitudes and Timing Central Meridian Transits

Measuring and Estimating Latitudes of Saturn's Global Features

One of the most important visual quantitative research programs for Saturn observers involves precise determination of latitudes of different belts on the globe of the planet. The measured latitude positions of these features may change to a small degree periodically, and observational evidence suggests that the widths of belts and zones vary with time.

Four basic methods are available for determining belt latitudes on Saturn's globe, requiring varying degrees of instrumental sophistication. First, measurement of latitudes from drawings of Saturn constitutes a rather large portion of the mass of data accumulated over the years by observers. Essential to the success of this method is the correct positioning of features relative to each other on standardized drawing blanks, as well as the selection of the appropriate form having the correct value of *B*, the Saturnicentric latitude of the Earth referred to the plane of the rings. This technique is easy to use and often quite accurate because experienced observers are capable of producing generally reliable sketches of Saturn.

Second, measurement of latitudes on high-resolution photographs of Saturn may produce good results compared with the preceding method, but the lack of detail captured on film, even with large instruments, somewhat reduces the usefulness of this procedure. Good-quality photographs generally show only one or more conspicuous belts, and their edges are typically indistinct due to the long exposures required or because of the grainy emulsion characteristics of the film. Consequently, positions of features on photographs are often difficult to measure with any degree of consistency.

Third, measurement of latitudes on CCD and webcam images of Saturn (as well as frames captured and processed from videotape) forms a rapidly increasing, more accurate alternative to utilizing photographs for this purpose. The best images taken with CCD cameras and webcams show a remarkable wealth of detail on Saturn, and belt edges are usually clearly visible. Indeed, observers who have the capability of imaging Saturn with CCDs or webcams are encouraged to submit their work for subsequent latitude measurements and analysis. Alternatively, those who choose to measure their own images and submit resultant latitude data are wholeheartedly invited to do so; this saves considerable time for section coordinators, like the author, who must otherwise measure a vast number of images, analyze the results, and prepare an apparition report for publication in a timely manner!

Fourth, measurement of Saturnian belt latitudes requires the use of a filar micrometer, but this method demands a very rigid mounting and a reliable, accurate clock drive with variable speed tracking adjustments. In addition to requiring a filar micrometer, which may be very difficult and expensive to obtain, it necessitates a relatively large aperture that produces a sizeable, sharp, and brilliant image of Saturn. Good viewing conditions are essential as well so that belt edges are clearly visible for accurate measurements. If an observer is fortunate enough to have a filar micrometer, this method is quite useful and accurate when viewing conditions permit.

Of these four methods, the first and third are the most frequently used. The next step after initial measurements are made is to compute the eccentric (mean), planetocentric, and planetographic latitudes of belt edges, as well as embedded belt and zone features, using the central meridian (CM) of Saturn as the standard longitudinal point of reference.

If X_n denotes the measured distance of the feature from the north end of the CM (on Saturn's limb) and if X_s is its distance from the south end (on the opposite limb of the planet's globe) of the CM, in arbitrary units, then by Equation 7.1,

$$y = \frac{1}{2}(X_s - X_n) \tag{7.1}$$

where the resultant value, y, is positive (+) for features or belt edges north of the center of the disk and negative (−) for those south of the center. Let r be the measured polar radius of Saturn in the same units as X_n, X_s, and y, and let R be the ratio of the equatorial radius of the planet to its polar radius, which is about 1.10 for the planet Saturn. B denotes the Saturnicentric latitude of the Earth, or the tilt of the axis of Saturn toward the Earth, as discussed earlier, and is found in an appropriate ephemeris. B' is the Saturnicentric latitude of the sun referred to the ring plane. The eccentric (mean) latitude, E, of the feature is computed by using Equation 7.2:

$$\tan B' = R \tan B \tag{7.2}$$

where B and B' are positive (+) when north, and Equation 7.3:

$$\sin (E - B') = \frac{y}{r}. \tag{7.3}$$

The planetocentric (Saturnicentric) latitude, C, of the same feature or belt edge is derived by using Equation 7.4:

$$\tan C = \tan \frac{E}{R}, \tag{7.4}$$

while the planetographic (Saturnigraphic) latitude, G, of the feature may be determined by using Equation 7.5:

$$\tan C = R \tan E. \tag{7.5}$$

The eccentric (mean) latitude, E, will always be near the arithmetic mean of the planetocentric and planetographic latitudes, and this numerical rule is illustrated by Equation 7.6:

$$\frac{C + G}{2} = E. \tag{7.6}$$

For many years, only planetocentric latitudes were computed for Saturn, chiefly due to the common belief that these numerical values were more relevant to the observer on Earth. Planetographic latitude is perhaps more familiar to the observer of Jupiter, but it is just as easily applied to Saturnian features by using the appropriate equations. The sine of the eccentric (mean) latitude of a Saturnian belt edge or feature is the fraction of the polar semidiameter of the planet when the Earth is overhead at the equator of the planet. Furthermore, the product of the cosine of the eccentric latitude and the equatorial radius of Saturn is the radius of rotation at that specific latitude. Because of potential scientific value, it is standard practice nowadays to compute all three latitudes for Saturn's belt edges and discrete features associated with belts and zones.

The four methods described above have advantages and disadvantages, and observers' preferences are largely driven by equipment and observing experience. A fifth technique, however, for determining Saturnian latitudes has been adopted by the ALPO Saturn Section in the last 30 years, sometimes referred to as the Haas technique, after the observer who introduced and perfected the method, Walter H. Haas, ALPO founder and director emeritus.

This fifth method is purely visual and is employed directly at the eyepiece, where it is only necessary to estimate the fraction of the polar radius subtended on the CM of the planet by the belt or feature whose latitude is desired. The value y is measured along the CM of Saturn as the distance from the center of the disk to the feature being observed (+ when north), divided by the distance, r, from the center of the disk to either the N or S limb. The center of the disk is usually located accurately by referring to the symmetry of the ring system, and it is obvious that y/r is equal to 1.0 at best. The ratio, y/r, is estimated to the nearest 0.01 whenever possible, and once this ratio has been determined visually, it is a simple procedure to compute the latitude of the feature using the equations introduced earlier. The ALPO Saturn Section has developed computer programs that enable quick computation of the various latitudes using estimated y/r values, and observers who want a copy of the program (available on IBM-format diskettes or via E-mail attachment) should contact the author.

Latitude estimates obtained by the visual technique are surprisingly reliable once the method is fully understood and used for several apparitions. A personal equation may be applied after observers master the visual procedure, making it possible to know how accurate individual results are when compared with that of others. The technique is very easy to use, and it is possible to generate a large number of accurate, reliable estimates in a short span of time. Furthermore, the method can be successfully applied to faint global features that frequently do not photograph well. To enhance visual contrast of those features being estimated, use of a W82 filter is highly recommended. Belt edges are sharpened with this filter, and bluish and reddish features can be distinguished better when they lie adjacent to one another.

The ratio, y/r, obtained in this visual procedure may be recorded on the standard observing forms provided by the ALPO Saturn Section for subsequent derivation of latitudes, along with intensity and colorimetric estimates, already presumably entered on the form.

Central Meridian Transit Timings

On occasion, discrete detail is visible in the belts and zones on the globe of Saturn, and such phenomena are similar in form although much less distinct than are features seen in the atmosphere of Jupiter. Projections or appendages from specific belts, sometimes spreading into extended festoons, or bright spots in the zones, comprise the most frequently recorded types of atmospheric phenomena on Saturn. Such fine detail is comparatively rare, and even when present, it is often accessible only to moderately large apertures. Thus, it is inadvisable for observers with instruments smaller than about 15.2 cm (6.0 in) to adopt the recording of such features visually as a sole observing program. Nonetheless, when particularly prominent and long-enduring spots or disturbances on the globe become visible or revealed on CCD or webcam images, CM transits are immensely important. Rotation rates at different belt or zone latitudes on Saturn are not established well enough due to the general lack of long-lived detail for transit timings, especially at higher latitudes.

As the planet rotates in a prograde fashion (W to E in the IAU sense), whereby features are progressively carried across the globe from right to left in the simply inverted view (from the Northern Hemisphere of the Earth), markings become situated on the CM at different times. It is of interest to record the time of each CM event precisely. Furthermore, it is virtually certain that Saturn rotates in at least two systems like Jupiter, although more objective, confirmed evidence of differential rotation at various latitudes is desired.

Any feature that can be followed for a month, even if the CM transit timings are off by a minute or so, can yield some valuable results. Research reveals that the rotation of the equatorial region of Saturn takes place in $10^h14^m13.0^s$, which means that 7 rotations of the globe occur in about 71.5^h, and further transit timings can be attempted for the same feature (if it persists) within about 3.0^d of the initial sighting. This rotation rate is fairly well established for short-lived phenomena, which seems to be a characteristic of Saturn, and this region is referred to as System I, which includes the north equatorial belt (NEB), south equatorial belt (SEB), and the equatorial zone (EZ), inclusive of the often ill-defined equatorial belt (EB).

Regions north or south of system I show rotation rates of $10^h38^m25.0^s$, and it is relatively easy to predict subsequent returns to the CM of features in about 3.0^d of the first transit after a period of some 74.3^h elapses. These areas comprise system II, in accordance with the nomenclature system adopted by the ALPO Saturn Section. Saturn's internal rotation rate of $10^h39^m24.0^s$ is based on the periodicity of radio emissions, and is called system III, but this radio rate has little importance for visual observers.

Concentrated efforts have been underway for quite a number of years to establish a good enough rotation rate for various latitudes on Saturn, all in an effort to prompt publication of comprehensive CM longitudes for Saturn in the *Astronomical Almanac* every year (see Chapter 5). Such information is already available for Jupiter in the same publication, but only system III data for Saturn are included. Organizations such as ALPO and the British Astronomical Association (BAA) now publish system I, II, and III data on their Web sites for the convenience of observers. These data allow prediction of subsequent returns of features to the CM of Saturn on any given night of observation if the phenomena last long enough for recovery. Tables in various ephemerides for system I, II, and III give longitudes of Saturn's geocentric CM for the illuminated (apparent) disk at 0.0^h UT for each day of the year. Incorporating corrections for phase, light time, and the planetocentric longitude of the Earth, system I longitudes were originally based on a precise sidereal rotation rate of 844.0°/day (or 10^{h}14^{m}13.0^s). Later, this value was changed to agree with the IAU rate of 844.3°/day (10^{h}14^{m}00.0^s). System I is used for features in the NEB, SEB, and EZ. System II, intended for the rest of the globe of Saturn excluding the north polar region (NPR) and south polar region (SPR), assumes a sidereal rotation rate of precisely 812.0°/day (or 10^{h}38^{m}25.4^s). For polar regions, mainly the SPR and NPR, the rotation rate is normally considered equal to system I. As before, the accuracy of these rates is governed by latitude-dependent factors that are not well established. Yet, longitudes derived from the tables should give conveniently small drift rates in most cases.

Although the proper choice of one of these two systems usually produces small drift rates for most short-lived features, rotation rates of Saturn's features are highly variable. Periods observed have ranged from 10^{h}02^m (at the equator) to 11^{h}03^m (57.0° planetocentric latitude). Thus, it is only coincidental that the recently discovered system III radio rotation rate of 10^{h}39^{m}22^s (810.8°/day) is close to that of system II.

Recall that there is convincing evidence for a multiplicity of rotation rates for Saturn, so observers should make every effort to obtain CM transit timings of long-lived features (combined with latitude estimates or measurements) so that the pattern of Saturn's atmospheric rotation and circulation can be better understood.

It is perhaps worthwhile to consider an example of the usage of ephemeris data just discussed to compute the CM value for system I, II, and III. Assume that the center of a bright spot in the south component of the south equatorial belt (SEBs) is observed to transit the CM at 09^{h}57^m UT on 2004 May 15^d. Since the SEBs falls in system I, we may extract the following values from an ephemeris (in this instance, the *A.L.P.O. Saturn CM longitudes*, available on the Web, were used):

System I CM @ 0^h UT on 2004 May 15^d		347.4°
+ Motion of system I CM	09^h	316.6
	50^m	29.3
	7^m	4.1
System I CM @ 09^{h}57^m UT on 2004 May 15^d		697.4°
Subtract 360°		−360.0
Solution		337.4°

Always subtract 360° from values greater than 360°, and it is customary to round off visual timings to the nearest whole degree. For convenience in deriving precise CM longitudes from an appropriate ephemeris for a specific date and time, Table 7.1 lists values for the motion of Saturn's CM in intervals of mean time for system I, II, and III. In addition, knowing the longitude of a feature on the planet's globe,

Table 7.1. Motion of Saturnian CM longitude in intervals of mean time

h	°	m	°	m	°
System I (NEBs, EZ, SEBn)					
1	35.2	10	5.9	1	0.6
2	70.4	20	11.7	2	1.2
3	105.5	30	17.6	3	1.8
4	140.7	40	23.5	4	2.3
5	175.9	50	29.3	5	2.9
6	211.1	60	35.2	6	3.5
7	246.3			7	4.1
8	281.4			8	4.7
9	316.6			9	5.3
10	351.8			10	5.9
System II (areas N or S of system I)					
1	33.8	10	5.6	1	0.6
2	67.7	20	11.3	2	1.1
3	101.5	30	16.9	3	1.7
4	135.3	40	22.6	4	2.3
5	169.2	50	28.2	5	2.8
6	203.0	60	33.8	6	3.4
7	236.8			7	3.9
8	270.7			8	4.5
9	304.5			9	5.1
10	338.3			10	5.6
System III (origin of radio emissions)					
1	33.8	10	5.6	1	0.6
2	67.6	20	11.3	2	1.1
3	101.3	30	16.9	3	1.7
4	135.1	40	22.5	4	2.3
5	168.9	50	28.2	5	2.8
6	202.7	60	33.8	6	3.4
7	236.5			7	3.9
8	270.3			8	4.5
9	304.0			9	5.1
10	337.8			10	5.6

observers may utilize published ephemerides and Table 7.1 to predict when it should next transit Saturn's CM. If a feature lasts long enough, successive CM transits are useful in determining whether it is drifting ahead or lagging behind in rotation in a particular belt or zone. As a rough rule of thumb, bear in mind that system I longitudes repeat in close to 3.0^d, while system II longitudes repeat in approximately the same interval.

The simplest procedure for timing CM transits is to estimate to the nearest whole minute the time when the feature is exactly midway between the E and W limbs of the planet. Times should always be given in Universal time (UT), obtained by listening to (or synchronizing an accurate digital watch to) WWV or CHU time signals or referring to an atomic clock. An even more accurate procedure involves making CM transit timings in the form of three separate estimates, as illustrated in *Figure 7.1*:

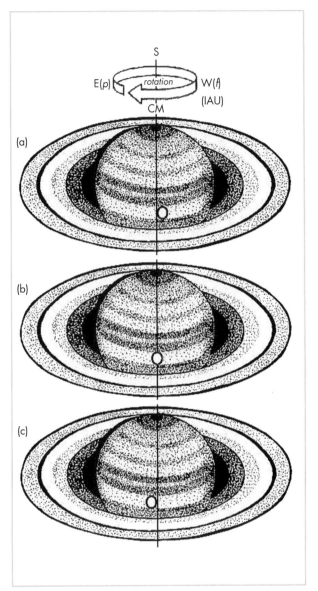

(a)

(b)

(c)

Figure 7.1. A hypothetical feature crossing the CM of Saturn as seen in a normal inverting telescope. Three timings are made as follows: (a) the UT when the features spot is precisely on the W or *f* side of the CM, (b) the UT when it is exactly centered on the CM, and (c) the UT when the feature is on the E or *p* side of the CM. Making three accurate estimates such as this greatly increases the value of CM transit timings. The direction of rotation and conventional IAU reference orientation appears in the diagram. (Credit: Julius L. Benton, Jr.; ALPO Saturn Section.)

1. Record the last minute (UT) when the feature's E edge is exactly on the west or following, *f*, side of Saturn's CM.
2. Record the last minute (UT) when the center of the feature is right on the CM.
3. Record last minute (UT) when the feature's W edge is precisely on the east or preceding, *p*, side of the planet's CM.

Always accompany transit timings of CM features with a numerical appraisal of the seeing and transparency conditions using the procedures discussed earlier in this book, instrumental factors, etc. It is critically important to include a description of the feature observed, the measured or estimated latitude of the marking (or the belt or zone in which it is located), and a sectional drawing to emphasize the morphology of the feature. Forms are available from the ALPO Saturn Section to facilitate effective recording of the above data. If the feature(s) being timed is captured on a digital image, make a separate copy of the image and insert an arrow, using a software program such as *Adobe Photoshop*, to point to the feature(s).

Once transit data are received by the ALPO Saturn Section for a particular feature, providing that a useful minimum number of transit timings are available, rotation periods of any individual spots or disturbances can be determined from drift charts. Only on rare occasions, however, does discrete detail appear on the globe of Saturn that persists long enough to allow derivation of rotation rates in a given latitude. When such recurring spots or disturbances appear on the planet, CM transits are extremely valuable.

Observers should refer to *Appendix A*, which contains a sample form for recording observations of CM transits of features detected on Saturn's globe. A blank appears on the form, adjacent to the space for recording CM transit data, for sectional sketches of the observed feature that is crossing the CM.

Observing Saturn's Satellites

Saturn is attended by at least 33 known satellites, eight of which are visible with telescopes normally available to amateur astronomers, but seeing them is considerably more problematic than expected due to their intrinsic faintness and relative proximity to the planet's brilliant globe and ring system. In addition, the changing inclination of the ring plane to our line of sight, as well as the varying distance of Saturn from the Earth, hampers visibility of the satellites. Using an eyepiece with a built-in occulting bar to block the globe and rings immensely improves the chances of detecting satellites in close proximity to the planet. When the rings disappear from view during edgewise presentations to our line of sight, which occurs at intervals of 13.75^y and 15.75^y (see Chapter 4), the most advantageous conditions occur for seeing Saturn's satellites.

In terms of visual magnitude, many of Saturn's satellites show small brightness amplitudes attributed to their varying orbital positions relative to the planet and because of the asymmetrical distribution of surface markings. Even after close proximity sensing by spacecraft, however, the true nature and extent of all of the observed satellite brightness variations is not completely clear and merits further study. Table 8.1 lists the satellites of Saturn that have any real importance to visual observers on Earth, including some fundamental data on the more accessible ones to instruments of moderate-to-large aperture.

Mimas is especially difficult to see except at greatest elongations (i.e., the maximum apparent distance of the satellites east or west of Saturn in the sky as

Table 8.1. Satellites of Saturn for visual observation

Designation and name of Saturn's brighter satellites	Average apparent diameter (")	Visual magnitude (m_v) at mean opposition	Amplitude
S1 Mimas	0.15	12.10	?
S2 Enceladus	0.13	11.77	?
S3 Tethys	0.28	10.27	0.25–0.50
S4 Dione	0.27	10.44	0.25–0.50
S5 Rhea	0.35	9.76	0.25–0.50
S6 Titan	0.70	9.39	0.24–0.60?
S7 Hyperion	0.10	14.16	?
S8 Iapetus	0.28	9.5–11.0	1.50–2.00

Estimating Satellite Magnitudes

seen from Earth), and although it is probably detectable with a 25.4 cm (10.0 in) telescope under optimum conditions, Mimas normally requires much more aperture to be clearly visible. Seeing Enceladus likely necessitates at least a 15.2 cm (6.0 in) telescope, while Tethys and Dione attain maximum brightness (just like Mimas) near greatest elongations and are frequently visible with 10.2 cm (4.0 in) instruments. Tethys is typically brighter at western as opposed to eastern elongation, while Dione is brighter at eastern elongation. Rhea is easy to see with a 7.5 cm (3.0 in) aperture, Titan is even visible in 7×50 binoculars if the observer knows its position relative to Saturn, while faint Hyperion might be within reach of a 15.2 cm (6.0 in) telescope. Maximum brightness of Iapetus occurs at western elongation—it is even brighter than Rhea at such times—and is visible in 7.5 cm (3.0 in) apertures. At eastern elongation, however, Iapetus diminishes so much in visual magnitude that a 15.2 cm (6.0 in) telescope is needed to see it. Of course, we now know from spacecraft flybys that the tremendous brightness differences associated with Iapetus occur largely because its lighter trailing face (best seen from Earth at western elongation) has extensive amounts of bright H_2O-ice, while its leading dark hemisphere (turned toward our view at eastern elongation) also has plentiful H_2O-ice but is overlain by dark carbonaceous deposits.

Estimating Satellite Magnitudes

Even though CCD photometry has virtually replaced visual magnitude estimates of the Saturnian satellites, visual observers can still do very useful work. The most reliable means for estimating satellite visual magnitudes at the eyepiece involves using standard stars of calibrated brightness for comparison when Saturn passes through a field of stars that have precisely known magnitudes. One possible source of such reliable reference star charts is the American Association of Variable Star Observers (AAVSO), but, unfortunately, Saturn only rarely crosses star fields that are covered by standard AAVSO star charts. Therefore, observers routinely utilize some of the more popular star atlases and catalogues that include faint stars and have reliable magnitudes for this kind of work. A number of excellent computer star atlases are gaining popularity and facilitate precise plotting of Saturn's path against background stars for comparative magnitude estimates. Another reference alternative involves Saturn's brightest satellite, Titan. It shows only subtle visual magnitude fluctuations over time compared with the other bright satellites of Saturn that have measured variances (see Table 8.1). Thus, since Titan's amplitude is not appreciable, it serves as a last-resort comparison standard at magnitude 8.4 when other reference objects are lacking.

Ideally, visual photometry of Saturn's satellites begins by first selecting at least two stars with well-established magnitudes and that have about the same color and brightness as the satellite. Make sure one of the stars is slightly fainter and the other is a little brighter than the satellite so that the difference in brightness between the stars is about 1.0 magnitude. This makes it easy to divide the brightness difference between the two comparison stars into equal magnitude steps of 0.1. To estimate the visual magnitude of the satellite, simply place it along the scale between the fainter and brighter comparison stars. For example, suppose that a given satellite appears only slightly fainter than a star, S_1, but quite a bit brighter than a star, S_2. If the satellite were, say, about 0.3 fainter than S_1, and hence 0.7 brighter than S_2, the observation takes the form

$$S_1(0.3)V_{os}S_2(0.7),$$

where V_{os} denotes the visual magnitude of the satellite, yet undetermined. If the magnitudes of stars S_1 and S_2 are 9.2 and 10.5, respectively (as determined from an appropriate star atlas and catalogue), derivation of the satellite's magnitude occurs as follows:

Given: Star S_1 has a visual magnitude, m_v, of 9.2
          Star S_2 has a visual magnitude, m_v, of 10.5
Find: $S_1 - S_2$ or $(9.2) - (10.5) = 1.3$
          $S_2 - S_1$ or $(10.5) - (9.2) = 1.3$
Find: Product of 1.3 and the fraction by which the satellite is fainter than S_1; that is, $(1.3)(0.3) = 0.39$
Add: 0.39 and the value of S_1; that is, $(0.39) + (9.2) = 9.59$ or 9.60 (rounding off)
Answer: 9.6 m_v (estimated visual magnitude of the satellite)

It is always important to keep careful notes in the observing record of all stars employed as reference objects, and be sure to identify the satellite in question precisely. Accompanying information, as recorded on the specific observing forms, must include the Universal time (UT) of the estimate, altitude of Saturn above the horizon, seeing and transparency, aperture and magnifications used, filters employed, and the location of the observing station. By following instructions outlined in publications such as the *Astronomical Almanac*, proper identification of the satellite in the eyepiece is relatively easy, and popular astronomical magazines like *Sky and Telescope* and *Astronomy* now publish finder charts that plot satellite positions accurately for the date and time in question. Satellite positions are also available from the ALPO and BAA Saturn sections, as well as on the Internet.

Some observers regularly use CCD cameras (with sufficient sensitivity) to image Saturn's satellites, along with any nearby comparison stars, as a permanent record to accompany visual observations and magnitude estimates. Images of the positions of satellites relative to Saturn on a given date and time are tremendously worthwhile for cross-checking against ephemeris predictions of their locations and identities (Fig. 8.1). It is important to realize, however, that the brightness of satellites and comparison stars captured on CCD images will not necessarily correspond to visual impressions because the peak wavelength response of the CCD chip is different from that of the eye.

Observers who have photoelectric photometers may contribute measurements of Saturn's satellites, but they are notoriously difficult to gauge owing to their faintness compared with the planet itself. Sophisticated techniques are required to correct for scattered light surrounding Saturn and its rings, although an exhaustive treatment of this subject is beyond the scope of this book. Those interested in this very specialized endeavor are encouraged to contact the Saturn sections of organizations such as the ALPO and BAA for further guidance and recommended contemporary literature.

Figure 8.1. Tofol Tobal of Barcelona, Spain, took this CCD image of Saturn's satellites Titan, Rhea, Dione, and nearby comparison star SAO 93454 (m_v = 7.05) on January 8, 2001, at 22:17 UT using a 10.2-cm (4.0-in) refractor. South is at the top and east toward the left in the normal inverted view. (Credit: Tofol Tobal; ALPO Saturn Section.)

Satellite Transits, Shadow Transits, Occultations, and Eclipses

When Saturn's rings are near or precisely at edgewise orientation (when the value of B is within $\pm 4.0°$ or smaller), the best opportunity exists for observing transits or shadow transits across the globe of Saturn of those satellites that lie near the planet's equatorial plane, much in the same way that such phenomena routinely occur for Jupiter's Galilean satellites. Occultations and eclipses by the globe of Saturn also happen at such times, as well as mutual phenomena of the satellites themselves. Instruments less than 30.5 cm (12.0 in) in aperture are usually insufficient to give good views of most of these phenomena for Saturn's satellites, except perhaps in the case of Titan, whose disk is roughly 0.6″ across when it transits Saturn's globe. Controversy persists as to whether shadow transits of any of the satellites other than Titan are visible from Earth with large instruments. The other satellites may simply be too small to cast umbral shadows onto the globe of Saturn.

Mutual phenomena, mostly involving Tethys, Dione, Rhea, and Titan, take the form of close passages or occultations of two or more satellites. For example, a pair of satellites may first approach one another, merge into an unresolved elongated image, and then subsequently move apart and appear as two separate bodies without ever passing through actual occultation (i.e., when one satellite is situated exactly in front of the other). Worth noting here are the fluctuations in brightness that take place. Close approaches that result in an unresolved pair produces a combined magnitude of both objects, so the apparent "single body" is brighter than the two individual satellites. When a true occultation occurs, where one satellite passes directly in front of the other, the combined magnitude diminishes at the point of superposition of one satellite upon the other. The brightness difference between the unresolved elongated image of the two satellites and that of the two satellites precisely in occultation is noticeable and worth recording. Brightness estimates occur relative to adjacent field stars of known visual magnitude or with other Saturnian satellites that are nearby but not participating in the mutual events. Capturing the progression of these interesting events on video tape or using digital cameras and webcams to get sequential images is especially meaningful.

Predictions of transits, shadow transits, occultations, and eclipses of Saturn's satellites, as well as mutual satellite phenomena, when the rings are at or near edgewise orientation commonly appear in specialized publications such as the *Journal of the ALPO* or are available from the Saturn sections of the ALPO or BAA, as well as the Internet. Because predictions can be off by a considerable amount, it is prudent to begin watching for satellite phenomena well in advance of the estimated time events are supposed to occur. Immediately dispatch to the ALPO or BAA Saturn sections precise timings (UT) to the nearest second of ingress, central meridian (CM) passage, and egress of a satellite or its shadow across the globe of Saturn at or near edgewise presentations of the rings. Remember to include the identity of the belt or zone on the planet crossed by the shadow or satellite. Intensity estimates of the satellite, its shadow, and the background belt or zone are useful as well, and drawings of the immediate area where the events occur at precise time intervals are especially useful. High-resolution digital images of these phenomena are vitally important, as well as videotape of the entire sequence of events. It is important to recall that edgewise ring presentations happen for only short periods during Saturn's 29.5^y orbital period, so observers will want to take advantage of such rare opportunities to record these satellite phenomena. At edgewise ring orientations, satellite magnitudes are easier to estimate because of the reduced glare, and brightness corrections are then generally only a function of the apparent distance of the satellite from the planet.

Visual observations of any markings on Saturn's satellites are beyond the scope of most amateur endeavors, except perhaps in the case of Titan. Extremely large instruments are essential, along with nearly perfect viewing conditions, for resolution of Titan as a reddish-orange disk. Even then, Titan does not offer up much because of its thick atmosphere, and it is therefore not surprising that few observers, using very large apertures, have ever submitted visual observations of any phenomena on Titan. Nevertheless, there are special professional–amateur projects that suitably equipped observers may participate in, and we discuss these now.

Specialized Observations of Titan

Titan is an extremely dynamic satellite exhibiting transient as well as long-term variations. From wavelengths of 3000 to 6000 Å, reddish methane (CH_4) haze in its atmosphere dominates Titan's color, while beyond 6000 Å, deeper CH_4 absorption bands appear in its spectrum. Between these CH_4 bands are "windows" to Titan's lower atmosphere and surface, and daily monitoring in these windows with photometers or spectrophotometers is worthwhile for cloud and surface studies. Saturn sections of the ALPO and BAA actively solicit regular spectroscopy of Titan as a meaningful professional–amateur cooperative project. Sporadic monitoring of

Titan occurs with large Earth-based instruments and is increasingly under scrutiny by spacecraft, and a definite need exists for good systematic observations with instrumentation available to amateurs, especially imaging with a CH_4 absorption filter. In addition, long-term investigations of other areas from one apparition to the next can help shed light on Titan's seasonal variations. Suitably equipped observers, therefore, are encouraged to participate in this interesting and valuable project.

Systematic long-term imaging of Titan in the infrared (IR) by amateur astronomers is another professional–amateur project that suitably equipped observers can actively participate in. If properly executed, this effort is of tremendous value as a supplement to Earth-based professional observations and spacecraft monitoring of the clouds on Titan. Observational sequences lasting a minimum of 1^h (but the longer the better) with telescopes of at least 31.8 cm (12.5 in) in aperture are recommended, while robotic telescopes are a great advantage for a program such as this, although not essential. The technique involves regular imaging of Titan, whenever weather permits, using short exposures of roughly 10^s with a pair of special narrowband pass IR filters. The recommended filter set is available in diameters of 12.5, 25.0, and 50.0 mm and transmit a well-defined bandwidth of 100 Å, one centered on 7500 Å and the other on 7947 Å, effectively rejecting other unwanted wavelengths. The only drawback is they are moderately expensive (in the range of $250 each), but they are essential for this type of work. One filter "sees" only the uppermost cloud layers of Titan, while the other penetrates much deeper into its atmosphere, and though the images are in the IR, the scenario is analogous to viewing planets at visible wavelengths; that is, some filters can penetrate down to the surface or various levels within the atmosphere and some do not. At IR wavelengths, the upper atmosphere of Titan does not change appreciably (or changes occur on time scales in excess of several years), so the ratio between the two brightness readings using these two filters immediately indicates whether or not the surface or deep cloud regions have changed in brightness. Excessive noise is usually present in single 10^s exposures, but imaging Titan for extended periods and afterward constructing a single brightness ratio from these data produces the best results. Because spacecraft only monitor Titan during a flyby, and large ground-based telescopes only look intermittently, having the capability of a team of observers watching nightly is a huge advantage. The greatest limitation of such a project, of course, is weather conditions on the Earth, which prevent opportunities for getting views on every single night. A large active network of amateur astronomers, however, provides a means for overcoming this problem and heightens the probability of getting a truly continuous record. A team spread out in longitude could look for variations on time scales of less than 1^d and help confirm suspicions that measurable deviations in Titan's simple light curve occur periodically. Imaging when some of these variations take place may aid in detection of cloud outbursts often suspected at the south pole of Titan. Suitably equipped advanced observers should definitely consider getting involved in this meaningful research project.

A Primer on Imaging Saturn and Its Ring System

Experience has shown that no matter how disciplined and well seasoned a visual observer is, the eye can be notoriously unreliable when interpreting planetary phenomena at or near the threshold of vision. Seasoned visual observers generally acknowledge the fact that it is nearly impossible to be completely objective in describing delicate contrasts and patterns seen or suspected on the surfaces and in the atmospheres of planets, and deducing the absolute color of different features. It is easy to misjudge what is seen in the eyepiece, which is one of the reasons why, as mentioned earlier in this book, simultaneous observations are so critically important for those who do mainly visual work. So, in addition to long-term systematic visual work described previously, which includes full-disk drawings, intensity estimates, central meridian (CM) transit timings, latitude measurements, and comprehensive descriptive reports, observers have periodically taken black and white as well as color photographs of Saturn using 35mm single-lens reflex (SLR) cameras in an attempt to reduce subjectivity in the data. Through trial-and-error efforts employing different types of film and a variety of photographic setups and techniques, conscientious amateur enthusiasts with a lot of patience have produced superb, high-resolution images of the planet (Fig. 9.1). Astrophotography of Saturn, nevertheless, always offers a momentous challenge because of the lack of clearly discernible features that show up on film, whereby most first-rate photographs of the planet depict perhaps only the south equatorial belt (SEB) or north equatorial belt (NEB), the equatorial zone (EZ), Cassini's division, the ring or globe shadow, and maybe a couple of the more conspicuous major ring components.

While higher resolution exposures of Saturn occasionally reveal some of the larger features within individual belts and zones (e.g., prominent bright and dark spots), particularly those made using color filters to enhance contrast and transmit different wavelengths of light, viewing conditions often preclude recording on film the finest details visible to keen-eyed visual observers. Another shortcoming of typical film photography is the necessity of enlarging planetary images by using eyepiece projection, which reduces light and renders images of Saturn very dim. This means longer exposures are required, even with extremely fast film, and higher powers amplify vibration of the telescope, ruining the sharpness of the photograph despite steady seeing. For this reason, many visual observers have

Astrovideography

Figure 9.1. Saturn was photographed on March 8, 1974, at 02:40 UT by Richard J. Wessling of Milford, Ohio using a 31.8-cm (12.5-in) Newtonian with Tri-X film in moderate viewing conditions. South is at the top and east is toward the left. The most obvious features in this photo are the EZ, SEB, dusky South polar hood, Cassini's division, Crape band, and the shadow of the globe on the rings. (Credit: Richard J. Wessling; ALPO Saturn Section.)

steadfastly maintained their practice of drawing Saturn as part of a simultaneous program instead of attempting 35mm photography because of frustrations associated with not being able to document on film their best visual impressions. Of course, high-quality photographs of Saturn by skilled individuals are always welcome and useful as supplements to a series of good visual observations for any apparition. An unbiased comparison of concurrent visual drawings and high-resolution 35mm photographs of Saturn helps shed light on the limits of visibility of elusive atmospheric or ring phenomena.

Astrovideography

Amateur astronomers have employed camcorders the last couple of decades to produce some very meaningful videos of Saturn, and they are especially advantageous for group observing endeavors. One of the wonderful rewards of astrovideography is the opportunity to record dynamic events, such as occultations of Saturn by the moon, transit of Titan's shadow across the globe, other phenomena near edgewise ring presentations, and passage of a bright star behind the rings. Other than the main telescope, the equipment and methodology required for getting favorable results is minimal, since the basic prerequisites are a camcorder, a VHS recorder, and a television monitor. The camcorder should have a removable lens, however, since there is no need for it when recording video frames through a telescope. Most home camcorders pick up about 30 images per second, which means in 5 minutes they produce 9000 frames, and the best results are usually obtained with camcorders featuring low lux numbers (<2.0) as well as provisions for automatic and manual focus.

Figure 9.2. Specialized video cameras for imaging the planets are both lightweight and relatively easy to use, like the extremely successful Astrovid Color PlanetCam shown here. (Credit: Adirondack Video Astronomy.)

As an alterative to using home camcorders, for a number of years some observers relied on run-of-the-mill security cameras with removable lenses for imaging Saturn. In the proper hands, models featuring manual shutters, adjustable gain, lux ratings between 0.1 (black and white cameras) and 1.0 (color cameras), and resolution better than 300 lines worked very well. Nowadays, far superior planetary video cameras have burst on the scene at very competitive prices, relegating generic surveillance cameras to near obsolescence for solar system imaging (Fig. 9.2). These unique planetary video cams run on DC or AC power and are considerably cheaper than most CCD cameras, plus most are remarkably lightweight. They attach to a telescope for prime focus astrovideography using a C-mount or similar adapter without significantly throwing off the balance of the instrument, and their NTSC composite video signals are routed via an industry-standard cable to a TV monitor or VHS recorder. This camera-to-telescope setup gives high magnifications, especially when focal ratios surpass f/10, so a clock-drive with slow motion controls and a finder with crosshairs are considered essential. With shorter focal length optical systems, a 2× or 3× Barlow lens helps increase the image size of Saturn. Color filters, and perhaps a variable density polarizer, enhance wavelength-dependent features and accentuate subtle contrast differences.

Planetary video cameras produce horizontal resolutions upward of 480 lines, necessitating the use of Super-VHS recorders that support these higher resolutions (standard VHS recorders are rated at only 240 TV lines), and matching color or

Figure 9.3. Available in color and monochrome models, battery-powered electronic imaging eyepieces, like the one shown here, neatly fit into the focusing tube of most telescopes and will transmit live video images of Saturn via a small RCA cable to a TV monitor for viewing or a VCR for recording. They contain CCD detectors like the chip in a video camcorder, and although they work fine for planets and the moon, they are not sensitive enough for imaging faint deep sky objects. (Credits: Orion Telescopes and Binoculars.)

monochrome monitors (those that display better than 450 lines) are essential for live high-resolution viewing. Digital video decks powered by rechargeable batteries are another option for making quality recordings.

In addition to planetary video cameras, a few manufacturers offer very compact battery-powered electronic imaging eyepieces weighing only a few ounces that fit nicely into standard focusing tubes of most telescopes (look at the example shown in Fig. 9.3). They transmit real-time video images of Saturn over a small RCA cable to a TV monitor or Super-VHS recorder and embody small CCD detectors that compare reasonably well in resolution and sensitivity to regular video camcorders. For example, the newest monochrome electronic oculars have a pixel count of 320 × 240, a manual contrast control, and a sensitivity of 1.2 lux, while color versions have 510 × 492 pixel arrays, a sensitivity of 2.5 lux, and manual thumb wheel for tweaking hue and contrast. Both are capable of generating 60 frames per second.

While recording video at the telescope, viewing will randomly fluctuate on most evenings, so some of the frames will as a rule look crisper and more detailed than the rest. To selectively pull the best frames off the resulting Super-VHS tape, "image-grabbing" is necessary using a computer-installed capture card, or alternatively an external device connected to a parallel or Universal serial Bus (USB) port, together with operational software to adjust brightness, contrast, and color balance in a variable-size preview screen. Reasonably priced plug-and-play PCMCIA video capture cards have appeared on the market in recent years, compatible with many planetary video cameras, and they permit display and subsequent capture, as well as preview, of video images on laptop computers. Once frames are stored on the computer's hard drive, the sharpest and highest-contrast images are stacked, processed, and fine-tuned using a variety of powerful graphics programs.

The World of CCD Imaging

In recent years, the increased availability of CCD, digital, and web cameras at affordable prices (especially the latter) now permits more efficient light detection and image acquisition, coupled with shorter exposure times, than are achievable with ordinary film emulsions, in essence totally revolutionizing the field of planetary photography. Thus, in addition to visual observations where the trained eye of an experienced observer may recognize fine planetary detail in superb viewing conditions, the CCD chip permits permanent documentation of those same details in momentary periods of excellent seeing. The enormous dynamic range of CCD cameras, for example, and their high sensitivity to visual and near-infrared (IR) wavelengths make them a valuable tool in lunar and planetary research. Furthermore, all of the aforementioned devices create digital information that can be stored on a laptop or desktop computer for subsequent manipulation by image processors and graphics programs.

Because digital imaging is now so commonplace in lunar and planetary astronomy and seems to be the wave of the future, with results far surpassing those attainable with normal photographic film (except perhaps color accuracy), there is no discussion in this book of techniques of 35mm planetary photography, now considered to be an outmoded methodology by a rapidly growing number of solar system enthusiasts. Instead, the focus here is on the more contemporary techniques of CCD imaging, at the same time stressing simplicity, affordability, and ease of use with respect to equipment offering the greatest probability of success for producing high-resolution planetary images. Accordingly, the main emphasis in this chapter is on using digital cameras and webcams for imaging Saturn including some of the basic techniques for processing of the resultant images.

CCD and Digital Cameras

For at least 20 years CCD cameras of increasing complexity have been available to amateur astronomers, and for the most part, commercially available detectors can capture 70% or more of the incident photons (i.e., they have good quantum efficiency) with no reciprocity failure commonly encountered in film astrophotography. Most CCD cameras (e.g., Fig. 9.4) attach to a telescope using eyepiece projection couplers and a C-mount adapter, and connection to a computer is usually mandatory due to the need for operational software. Some of the most popular astronomical magazines are replete with glossy advertisements and accompanying images implying fantastic results are possible with CCD camera arrays right out of the box. Although they are, indeed, quite capable of producing high-resolution images like the wonderful example in Figure 9.5, the reality is that a great deal of expertise is usually necessary to obtain truly consistent superior results. Choosing the right CCD camera for lunar and planetary work is often quite a challenge, while getting the best images requires a wealth of time, trial, and effort. The better CCD

Figure 9.4. A very impressive CCD array coupled to a 31.8-cm (12.5-in) Newtonian owned by Eric Ng of Hong Kong. (Credit: Eric Ng; ALPO Saturn Section.)

cameras used for planetary work are generally quite expensive, which limits the number of observers who actually acquire and use them.

Digital cameras, which use a CCD chip to capture images, have become quite popular for use in lunar and planetary astrophotography in recent years. They eliminate the need for purchasing and subsequently developing photographic film, plus images are visible right away at the telescope on the camera's viewing screen. Most digital cameras have removable memory cards of varying capacity, permitting storage of many more images than the number of exposures on a roll of 35mm photographic film. Since they do not utilize chilled chips like bona fide CCD cameras, short exposure times are the rule to avoid imaging-degrading noise, but this is not a problem with bright solar system objects like the moon and planets,

Figure 9.5. Ed Grafton of Houston, Texas, using a 35.6-cm (14.0-in) SCT and a ST5c CCD camera, made this extremely detailed colorful CCD image of Saturn on January 5, 2003, at 05:30 UT. A wealth of detail is visible on Saturn's globe and in the rings, surpassing some of the best photographs of Saturn taken with relatively large instruments using conventional film. South is at the top and east toward the left. (Credit: Ed Grafton; ALPO Saturn Section.)

which can be successfully imaged in only a couple of seconds. Once a pool of images is stored within the camera, they only need to be downloaded over a USB cable to a laptop or desktop computer for processing using specialized software.

Lightweight high-resolution 4.1-to-6.0 megapixel digital cameras, powered by lithium or rechargeable Ni-Cd (nickel-cadmium) batteries, are now readily available at prices under $500. As rapidly as technology changes these days, more sophisticated models become available almost monthly, with prices for so-called superseded models dropping considerably, so bargains appear from time to time. The image quality that a given digital camera delivers is dependent on the number of pixels on the CCD chip, where more pixels mean finer details in the captured image. Digital cameras do not normally have lenses that are removable, which necessitates afocal imaging of planets directly through the eyepiece of the telescope. The most popular ones used for planetary work have filter threads on their lenses that accept readily available adapters for direct coupling to the telescope without having to hold the camera in hand at the eyepiece (Fig. 9.6).

Nearly all digital cameras have a zoom function for increasing magnification, but one that zooms internally instead of externally is almost mandatory for planetary imaging. Provisions for noise reduction and a remote shutter function to lessen vibration are always useful. Akin to eyepiece projection used with 35mm photography, these couplers enable precise centering of the camera along the optical axis and in turn position it at just the right distance from the top of the eyepiece to minimize vignetting. This cropping effect occurs when the cone of light emanating from the ocular fails to fully illuminate the CCD chip, and ideally the lens of the digital camera should be slightly smaller in diameter than the external lens of the eyepiece. Preferable oculars must have decent eye relief, flush-mounted

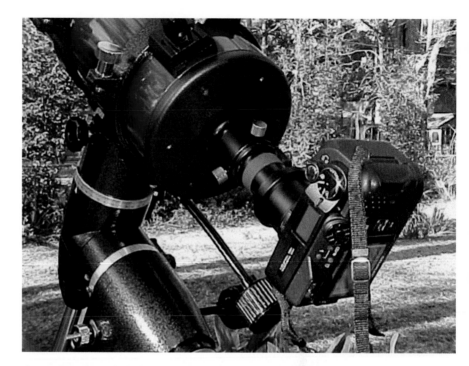

Figure 9.6. The author's high-resolution 4.1-megapixel digital camera, shown here, conveniently attaches to the focusing tube of the telescope using a special adapter that couples to both the eyepiece and lens housing of the camera. (Credit: Julius L. Benton, Jr.)

eye lenses, and shorter focal lengths, and avoid those that have recessed lenses and large rubber eyecups.

Precise focusing is essential to get good, crisp images, a process that can be tricky unless the LCD viewing screen of the camera is sizeable enough, but the afocal method generally means that images of the planets will be quite large, so this is not as troublesome at it may seem. The viewing screen also helps the observer manually adjust exposures for bright objects like planets. When using increasingly shorter focal length eyepieces, image size will proportionately increase, but so will vignetting and difficulties encountered with focusing correctly, let alone vibration, while racking the focuser in and out. It may be preferable to use 2× or 3× Barlow lenses in conjunction with longer focal length eyepieces having excellent eye relief, but quite a bit of experimentation is needed to achieve the correct overall combination.

Other than disabling the flash mechanism of the camera, most observers leave automatic settings alone. Sometimes higher International Standards Organization (ISO) settings enable shorter exposures and improved image sharpness, but this also can introduce unwanted noise if used indiscriminately. To get sharp images, precise focusing is mandatory. Most of the time, observers leave the digital camera in macro mode, while the telescope is relied on for the focusing function.

In some instances, the raw image produced by a digital camera may look perfectly acceptable, but most of the time stacking several images is a prerequisite for bringing out subtle detail and simultaneously reduce noise. It makes sense to stack and subsequently process only the best images. Many times, the software bundled with the digital camera accommodates resizing and cropping of images, as well as allowing adjustment of color balance and contrast, but specialized software is required to convert file types, stack images, and enable further enhancements.

Webcams

Webcams, originally developed for video conferencing over the Internet, have found a welcome home in lunar and planetary astronomy. Quite a few devout CCD

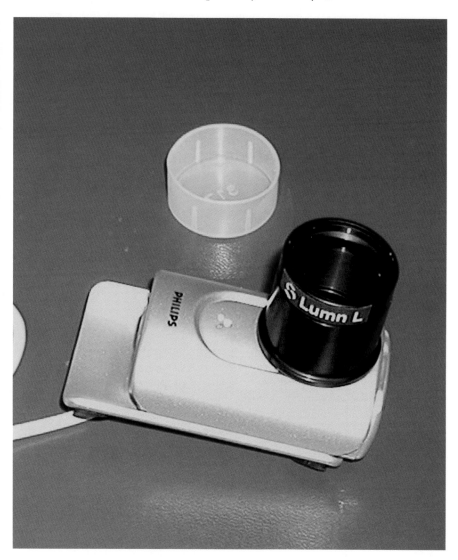

Figure 9.7. The author's webcam pictured here is extremely lightweight, connects to a laptop computer via a USB cable, and comes with an adapter that slips into the drawtube of a telescope's focusing mechanism. An IR blocking filter, which is highly recommended (see text), threads easily onto to the webcam. (Credit: Julius L. Benton, Jr.)

camera users, who first voiced skepticism when hearing that these little devices could rival more expensive CCD cameras, later wound up changing their minds when they saw the resulting images. Without a doubt, webcams are swiftly becoming the preferred CCD devices for imaging the planets. They are very easy to use and inexpensive, with prices starting just under $150. Webcams are also extremely lightweight and compact, and connect to a computer by way of an available USB port, and those sold for astronomy purposes feature removable lenses and convenient adapters that easily slip into the eyepiece holder of the telescope, as seen in Figure 9.7. To make them suitably within the means of most amateur astronomers, webcams embody relatively small CCD chips (preferred models use CCD chips as opposed to CMOS chips), yet their smaller pixels mean higher resolution and relatively large images. Viewing of images essentially occurs in real time on a computer monitor, and focusing takes place by merely looking at the computer display while adjusting the rack and pinion mechanism of the telescope (Fig. 9.8). Software on the computer to which the webcam is attached manually or automatically controls gain, brightness, shutter speed, gamma, and color saturation.

Unlike conventional CCD cameras, webcams can generate a huge number of image frames over the course of an observing session. Normal production webcams cannot handle exposures in excess of about 2 seconds, but this kind of sensitivity is perfectly fine for lunar and planetary work where long exposures are seldom required. Therefore, since brief exposures produce a multitude of raw

Figure 9.8. Paul Maxson of Phoenix, Arizona, uses this 23.5-cm (9.25-in) SCT and webcam setup to regularly image Saturn and other solar system objects. (Credit: Paul Maxson; ALPO Saturn Section.)

images, the probability of getting frames captured in moments of better-than-average viewing is higher. Although it may seem desirable to secure the maximum possible number of frames per second, in practice the limited transfer rate of the USB port necessitates data compression, which degrades image quality. To avoid this problem and get the sharpest images, observers seldom exceed frame rates of 5 to 10 per second. Most webcams have no provision for cooling like their CCD camera counterparts, but noise is largely imperceptible with the short exposures typically encountered with brighter planets like Saturn. Later, after the observing session ends, the best frames are selected from the raw data, and image processing on a computer can begin to adjust color balance and sharpness, brightness, and contrast. Ultimately, the highest resolution frames are stacked and manipulated using software programs, and color filters are not mandatory for getting color webcam images of the planets. Excessive processing, however, will unfortunately enhance noise, so tweaking the image should cease after achieving the best combination of all factors and a natural, realistic looking image appears on the computer screen. Because specialized software is required to capture images and subsequently process the raw data, it is necessary to download free programs from the Internet or purchase them (I am always willing to offer suggestions as to the most current ones available).

Procedure for Capturing and Processing Webcam Images

A comprehensive treatment of how to acquire and process digital images by various devices is beyond the scope of this book, but the latest information on this extremely dynamic subject is readily accessible on the Internet and in specialized astronomy literature. Since webcams have virtually supplanted all other means of CCD imaging by Saturn observers, it seems relevant to expand a little on the fundamental techniques of capturing images and stacking frames all in an effort to produce scientifically useful representations. The techniques described here apply to most of the popular webcams in use today by those who image Saturn.

As mentioned earlier, webcams are so convenient to use and are so affordable that getting started in this form of CCD imaging is simple. Assuming that a laptop computer has been powered up and situated in close proximity to the telescope, the first step in imaging Saturn with a webcam is to locate the planet using a relatively low-power eyepiece, then bring the planet to a precise focus. Turn on the telescope's clock drive, since the polar axis should already be aligned with the north celestial pole (NCP), and replace the eyepiece with the webcam. A live 24-bit color video image of Saturn should appear on the screen of the computer monitor. Use the operating software to set the size of the viewing window to 640 × 480 pixels. If necessary, adjust the slow-motion controls of the equatorial mounting to center the image in the display window, and then fine-tune the focus until Saturn and its ring system appear sharp (some observers utilize an eyepiece that is parfocal with the webcam to establish a focal point frame of reference). When imaging at focal ratios exceeding f/20, it becomes progressively challenging to get the planetary image centered on the 640 × 480 display window. A convenient accessory—a flip mirror assembly—helps solve this problem nicely. It is attached between the focusing back or drawtube of the telescope and the webcam. The webcam remains situated along the optical axis of the telescope, while an eyepiece is positioned at right angles to the array comparable to a star diagonal. Inside this accessory is a small mirror that is flipped up and down using a small knob, which directs the light cone into a crosshair eyepiece for proper focusing and centering in the field of view, then alternately into the webcam for imaging. By experimentation, a nearly parfocal setup for the eyepiece and webcam is achievable.

Focusing is not usually a major issue when viewing is steady, but things become more troublesome when the atmosphere is turbulent. Just like when making visual observations and drawings, selecting a night for using a webcam when viewing is good means fewer focusing issues and a greater opportunity for getting a large number of crisp raw images. Also, CCD chips are exceptionally sensitive to infrared (IR) wavelengths, which come to focus at a slightly different position than those of visible light. Consequently, images often appear fuzzy, especially in refractors (even apochromatic systems) and most catadioptrics. The remedy for this problem involves using an IR blocking filter that threads onto the adapter that attaches the webcam to the telescope, just like any other eyepiece filter.

Use the imaging software options of the computer program to turn off video compression. Although exposure times vary with focal length, try 1/25th second for Saturn with focal ratios of f/20 and greater, and start off by setting the frame rate to 10 fps (frames per second), brightness and contrast each to 50%, gamma to 20%, and gain to 40%. Once the shutter speed is at 1/25th second, adjust the gain so the planet's image in the display window appears a bit underexposed. Enter the duration for capturing raw images to between 60 and 180 seconds. Lastly, before closing the options menu, save the aforementioned settings!

Initially, the webcam takes a sizeable number of images of Saturn for the specified period of time, which will be saved as *.avi files on the computer. Getting a fairly high number of images means the probability is higher that some were made during moments of very good viewing. Of course, when viewing is better than average to start with, the higher the number of good images to work with. Be aware that a recording interval of 60 seconds at 10 fps generates 600 images of Saturn and a file size upward of 0.5 GB, so expect to consume a fair amount of disk space on the computer's hard drive during any given observing session. Once the raw images are captured by the webcam and transferred over a USB cable to the computer, it is wise to copy the files to a rewritable CD, DVD, or high-capacity USB "jump drive" for protection in the event a computer crash occurs or if files somehow become corrupted or accidentally deleted.

Do not expect separate frames to appear overly impressive, but browse through the raw images and select one that appears sharper and exhibits more detail than the rest. The processing software permits alignment of a sequence of raw images relative to this reference frame, and it also positions frames on the basis of quality with respect to one another. It also automatically stacks the best raw images, reduces noise, and enhances the quality of the final image. Stacking a few hundred of the very best frames measurably improves the signal-to-noise ratio (SNR). After completing the stacking process, which normally results in a good composite image, use a graphics programs to manipulate sharpness, color balance, and contrast. It is also possible to create interesting time-lapse animations of the rotation of Saturn, especially useful for showing motion of white spots and other discrete phenomena on the globe.

Figure 9.9. Thierry Lépine of Saint Etienne, France, imaged Saturn on February 9, 2005, at 21:45 UT using a 35.6-cm (14.0-in) SCT, a Philips ToUcam Pro webcam at f/27, and an IR blocking filter. The exposure time was 1/25th second and 1100 frames were stacked to produce this very detailed image. South is at the top and East toward the left. (Credit: Thierry Lépine; ALPO Saturn Section.)

Systematic Imaging of Saturn and its Ring System

In the visible part of the spectrum, there is very little absorption of electromagnetic radiation by the Earth's atmosphere, so imaging Saturn with CCD arrays, digital cameras, and webcams at visual wavelengths produces extremely impressive results showing numerous belts, zones, and associated detail on the globe and in the ring system components. Consider, for instance, some of the remarkably detailed images of Saturn presented in Figures 9.9 through 9.12 captured by skilled observers using various apertures with contemporary CCD imagers and webcams.

It is always meaningful to compare images of Saturn made by amateurs using different apertures, CCD cameras, and filter techniques to understand the level of detail seen, including any correlation with spacecraft imaging and results from professional observatories, and finally, how they relate to visual impressions of the globe and rings described in Chapter 4. Accordingly, in addition to routine visual studies, Saturn observers should carefully and systematically image the planet every possible clear night to search for individual features on the globe and in the rings, their motion and morphology (including changes in intensity and hue), to serve as input for combination with images taken by professional ground-based observatories and spacecraft monitoring Saturn at close range. Furthermore, comparing images captured over several apparitions for a given hemisphere of Saturn's globe provides information on seasonal changes long suspected by observers making visual numerical relative intensity estimates. It is noteworthy that images (and systematic visual observations) by amateurs occasionally serve as initial alerts of interesting large-scale features on Saturn that professionals may not already know about but can subsequently examine further with larger specialized instrumentation.

Figure 9.10. Notice all of the features on the globe and rings revealed in this amazing image of Saturn taken by Jason P. Hatton of Mill Valley, California, employing a 23.5-cm (9.25-in) SCT, a 3× Barlow lens, and a Philips ToUcam Pro webcam outfitted with a W25 (red) filter. Exposure time was 1/25th second and 1049 frames were stacked during processing. South is at the top and east toward the left. (Credit: Jason P. Hatton; ALPO Saturn Section.)

Particles in Saturn's atmosphere reflect different wavelengths of light in very distinct ways, which cause some belts and zones to appear especially prominent, while others look extremely dark, and imaging the planet using a series of color filters can help shed light on the dynamics, structure, and composition of its atmosphere. In the ultraviolet (UV) and IR regions of the electromagnetic spectrum, it is possible to determine additional properties as well as the sizes of aerosols present in different atmospheric layers not otherwise accessible at visual wavelengths, as well as useful data about the cloud-covered satellite Titan. UV

Figure 9.11. Toshihiko Ikemura of Nagoya, Japan stacked 1790 webcam frames to produce this beautiful color-balanced image of Saturn on November 7, 2003, at 17:01 UT using a 31.0-cm (12.2-in) Newtonian. Near the central meridian (CM) of Saturn is a small dark spot captured in the image in the south equatorial belt (SEB). (Credit: Toshihiko Ikemura; ALPO Saturn Section.)

Figure 9.12. From Hong Kong on October 18, 2003, at 20:26 UT, Eric Ng used his 31.8-cm (12.5-cm) Newtonian and a Philips ToUcam Pro webcam and 1/25th second exposures of Saturn at f/34.5 to render the colorful image; 800 frames were stacked during processing. South is at the top and east toward the left. (Credit: Eric Ng; ALPO Saturn Section.)

wavelengths shorter than 3200 Å are effectively blocked by the Earth's stratospheric ozone (O_3), while H_2O-vapor and CO_2 molecules absorb in the IR region beyond 7270 Å, and the human eye is insensitive to UV light short of 3200 Å and can detect only about 1.0% at 6900 Å and 0.01% at 7500 Å in the IR (beyond 7500 Å visual

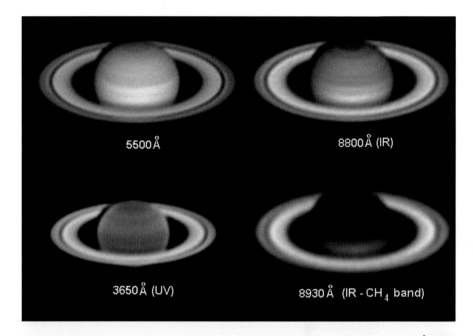

Figure 9.13. Shown here are four CCD images of Saturn taken at wavelengths of 5500 Å (RGB-visual), 3650 Å (UV), 8800 Å (near-IR), and 8930 Å (IR CH₄ band). South is toward the top and east is to the left in this normally inverted view as seen in most astronomical telescopes. The UV image was taken by Christophe Pellier of Bruz, France, at 03:43 to 04:29 UT on November 25, 2004, using a 21.0-cm (8.3-in) Cassegrain. The other three images were made from Japan by Tomio Akutsu on December 7, 2000, between 15:07 UT and 15:31 UT using a 65.0-cm (25.6-in) Cassegrain. Note how Saturn's appearance varies with wavelength band. Especially apparent is the contrast reversal in the IR CH₄-band, a spectral region where light is absorbed by CH₄, while the EZ is bright because high clouds there reflect IR back into space before being absorbed by CH₄. The highest altitude clouds, which tend to absorb UV light, appear dark in the UV image. (Credits: Christophe Pellier and Tomio Akutsu; ALPO Saturn Section.)

sensitivity is essentially nil). Although most of the reflected light from Saturn reaching terrestrial observers is in the form of visible light, some UV and IR wavelengths that lie on either side and in close proximity to the visual region penetrate to the Earth's surface, and capturing images of Saturn in these near-IR and near-UV bands frequently produces very interesting results (Fig. 9.13). For example, the effects of absorption and scattering of light by the planet's atmospheric gases and clouds of various heights and thicknesses are usually noticeably apparent, and such images sometimes show differential light absorption by particles with dissimilar hues intermixed with Saturn's white NH₃ clouds.

Saturn's atmospheric gases efficiently scatter light from the sun at shorter visual wavelengths, but this effect is much more evident in the UV region where gaseous H₂ and He disperses UV light so strongly, making the atmosphere look bright. Only the highest altitude cloud particles, which tend to absorb UV light, will appear dark against the bright background. In blue wavelengths scattering of light by atmospheric gases is not as pronounced as it is in the UV, so sunlight penetrates farther down into Saturn's cloud layers before some of it is reflected back toward the observer on Earth. Higher clouds in the equatorial regions, which are very reflective at visible wavelengths, are more obvious in blue light. The rings usually appear dark in the UV because they inherently reflect little light in this region.

At certain visible and IR wavelengths, light absorption by CH_4 blocks all but the uppermost layers of Saturn's atmosphere, which helps in the detection of clouds at different altitudes; that is, a reversal in contrast occurs in images taken in spectral regions where light is absorbed by CH_4 gas but scattered by high clouds. The EZ in near-IR images often looks bright because the high clouds there reflect long wavelength light back into space before much of it can be absorbed by CH_4.

Amateur imaging of Saturn in the near-IR or near-UV is a valuable complement to professional studies involving specialized instruments placed onboard spacecraft to monitor IR and UV wavelength regions that are not readily detectable from ground-based observatories (Fig. 9.14). Spacecraft investigations in the IR show that giant planets like Jupiter, Saturn, and Neptune not only reflect thermal energy from the sun, but radiate some of their own internal heat as well, and similarly, remote sensing by space probes in the UV region reveal aurora displays on both Jupiter and Saturn.

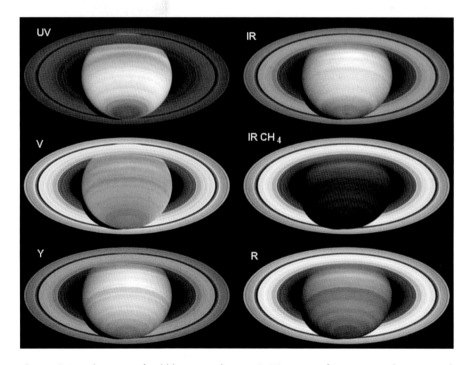

Figure 9.14. This series of Hubble space telescope (HST) images of Saturn was taken on March 7, 2003, at visual wavelengths (violet, yellow, and red) as well as in IR and UV regions. Different wavelengths of light reveal important characteristics of Saturn's atmosphere because cloud particles reflect different wavelengths in discrete ways, causing some features to stand out vividly while others are darker. Notice also the change in appearance of Saturn's rings at different wavelengths. North is at the top in these HST images. (Credit: NASA and E. Karkoschka, University of Arizona.)

Appendix A: Association of Lunar and Planetary Observers Forms

Association of Lunar and Planetary Observers (ALPO) Saturn Section: Central Meridian (CM) Transit Data and Sectional Sketches
(Attach this form to main observation form)

Observer: _____

Object: _____

UT Date:_____ UT Time:_____	
Location:_____(do sectional drawing at right)	
CM I: _____° CM II: _____° d_1: _____ d_2: _____	

S

p ☐ d_2 f

d_1

N

Object: _____

UT Date:_____ UT Time:_____	
Location:_____(do sectional drawing at right)	
CM I: _____° CM II: _____° d_1: _____ d_2: _____	

S

p ☐ d_2 f

d_1

N

Object: _____

UT Date:_____ UT Time:_____	
Location:_____(do sectional drawing at right)	
CM I: _____° CM II: _____° d_1: _____ d_2: _____	

S

p ☐ d_2 f

d_1

N

Object: _____

UT Date:_____ UT Time:_____	
Location:_____(do sectional drawing at right)	
CM I: _____° CM II: _____° d_1: _____ d_2: _____	

S

p ☐ d_2 f

d_1

N

Sectional Sketch Notation: d_1 = longitudinal extent in arc sec ( ) p = preceding
(all directions are IAU) d_2 = latitudinal extent in arc sec ( ) f = following

Association of Lunar and Planetary Observers (ALPO) Saturn Section: Visual Observation of Saturn's Satellites

(Attach this observation form to the main observing form for the same observing date)

Observer: _____   UT Date: _____

Reference Used for Locating Satellites:_____

Basic Symbolism Employed:

V_{os}	=	Visual magnitude of Saturn's satellite (computed from estimate)
X	=	Magnitude of comparison star (brighter reference star)
Y	=	Magnitude of comparison star (dimmer reference star)
>	=	Brighter than
<	=	Dimmer than

Note : All magnitudes are visual magnitudes derived from a reliable star catalogue for comparison stars

Satellite (name)	Comparison Stars Utilized in Estimates						Magnitude Estimates for Satellites		
	Star X			Star Y			Tenths < X	V_{os}	Tenths > Y
	Designation:			*Designation:*					
	m_v	RA	DEC	m_v	RA	DEC			

Source utilized for comparison stars:

Descriptive notes:

Association of Lunar and Planetary Observers (ALPO) Saturn Section: Visual Observation of Saturn for *B* = 0° (Edgewise Rings) Through *B* = ± 4°

(Rings are always omitted on drawing blanks for values of B ≤ 4°)

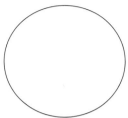

Coordinates (check one): [ ] IAU   [ ] Sky

Observer _____ Location _____

UT Date (start) _____ UT Start _____ CM I (start) _____° CM II (start) _____° CM III (start) _____°

UT Date (end) _____ UT End _____ CM I (end) _____° CM II (end) _____° CM III (end) _____°

B = _____° B = _____° Instrument _____ Magnification(s) _____ Xmin _____ Xmax

Filter(s) IL(none) _____ f₁ _____ f₂ _____ f₃ _____ Seeing _____ Transparency _____

Saturn Global and Ring Features	Visual Photometry and Colorimetry IL	f₁	f₂	f₃	Absolute Color Estimates	Latitude Estimates ratio y/r

Bicolored Aspect of the Rings: (always use IAU directions)

No Filter (IL) (check one): [ ] E ansa = W ansa [ ] E ansa > W ansa [ ] W ansa > E ansa
Blue Filter (_____) (check one): [ ] E ansa = W ansa [ ] E ansa > W ansa [ ] W ansa > E ansa
Red Filter (_____) (check one): [ ] E ansa = W ansa [ ] E ansa > W ansa [ ] W ansa > E ansa

Note: Attach to this form all descriptions of morphology of atmospheric detail, as well as other supporting information. Please do not write on the back of this sheet. The intensity scale employed is the Standard *ALPO Intensity Scale*, where 0.0 = completely black ⇔ 10.0 = very brightest features, and intermediate values are assigned along the scale to account for observed intensity of features.

Association of Lunar and Planetary Observers (ALPO) Saturn Section: Visual Observation of Saturn for *B* = ± 6° to ± 8°

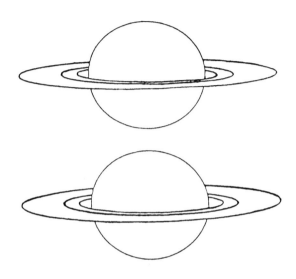

Coordinates (check one): [ ] IAU [ ] Sky

Observer _____ Location _____

UT Date (start) _____ UT Start _____ CM I (start) _____ ° CM II (start) _____ ° CM III (start) _____ °

UT Date (end) _____ UT End _____ CM I (end) _____ ° CM II (end) _____ ° CM III (end) _____ °

B = _____ ° B = _____ ° Instrument _____ Magnification(s) _____ X$_{min}$ _____ X$_{max}$

Filter(s) IL(none) _____ f$_1$ _____ f$_2$ _____ f$_3$ _____ Seeing _____ Transparency _____

Saturn Global and Ring Features	Visual Photometry and Colorimetry				Absolute Color Estimates	Latitude Estimates ratio y/r
	IL	f$_1$	f$_2$	f$_3$		

Bicolored Aspect of the Rings: No Filter (IL) *(check one):* [ ] E ansa = W ansa [ ] E ansa > W ansa [ ] W ansa > E ansa
(always use IAU directions) Blue Filter (_____) *(check one):* [ ] E ansa = W ansa [ ] E ansa > W ansa [ ] W ansa > E ansa
Red Filter (_____) *(check one):* [ ] E ansa = W ansa [ ] E ansa > W ansa [ ] W ansa > E ansa

Note: Attach to this form all descriptions of morphology of atmospheric detail, as well as other supporting information. Please <u>do not</u> write on the back of this sheet. The intensity scale employed is the Standard *ALPO Intensity Scale*, where 0.0 = completely black ⇔10.0 = very brightest features, and intermediate values are assigned along the scale to account for observed intensity of features.

Association of Lunar and Planetary Observers (ALPO) Saturn Section: Visual Observation of Saturn for $B = \pm 10°$ to $\pm 12°$

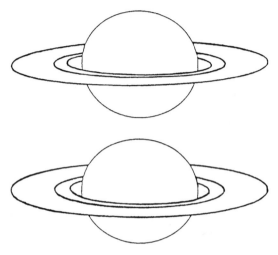

Coordinates *(check one)*: [ ] IAU   [ ] Sky

Observer _____ Location _____

UT Date (start) _____ UT Start _____ CM I (start) _____° CM II (start) _____° CM III (start) _____°

UT Date (end) _____ UT End _____ CM I (end) _____° CM II (end) _____° CM III (end) _____°

B = _____ ° B = _____ ° Instrument _____ Magnification(s) _____ x$_{min}$ _____ x$_{max}$

Filter(s) IL(none) _____ f$_1$ _____ f$_2$ _____ f$_3$ _____ Seeing _____ Transparency _____

Saturn Global and Ring Features	Visual Photometry and Colorimetry				Absolute Color Estimates	Latitude Estimates ratio y/r
	IL	f$_1$	f$_2$	f$_3$		

Bicolored Aspect of the Rings: No Filter (IL   ) *(check one)*: [ ] E ansa = W ansa [ ] E ansa > W ansa [ ] W ansa > E ansa
(always use IAU directions) Blue Filter (_____) *(check one)*: [ ] E ansa = W ansa [ ] E ansa > W ansa [ ] W ansa > E ansa
          Red Filter (_____) *(check one)*: [ ] E ansa = W ansa [ ] E ansa > W ansa [ ] W ansa > E ansa

Note: Attach to this form all descriptions of morphology of atmospheric detail, as well as other supporting information. Please <u>do not</u> write on the back of this sheet. The intensity scale employed is the *Standard ALPO Intensity Scale*, where 0.0 = completely black ⇔ 10.0 = very brightest features, and intermediate values are assigned along the scale to account for observed intensity of features.

Association of Lunar and Planetary Observers (ALPO) Saturn Section: Visual Observation of Saturn for B = ± 14° to ± 16°

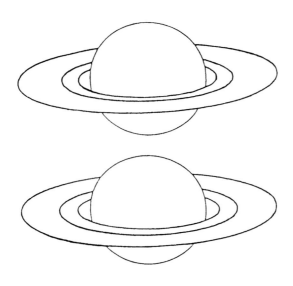

Coordinates (check one): [ ] IAU [ ] Sky

Observer _____ Location _____

UT Date (start) _____ UT Start _____ CM I (start) _____° CM II (start) _____° CM III (start) _____°

UT Date (end) _____ UT End _____ CM I (end) _____° CM II (end) _____° CM III (end) _____°

B = _____° B = _____° Instrument _____ Magnification(s) _____ x_min _____ x_max

Filter(s) IL(none) _____ f₁ _____ f₂ _____ f₃ _____ Seeing _____ Transparency _____

Saturn Global and Ring Features	Visual Photometry and Colorimetry IL	f₁	f₂	f₃	Absolute Color Estimates	Latitude Estimates ratio y/r

Bicolored Aspect of the Rings: *(always use IAU directions)*
No Filter (IL) *(check one):* [ ] E ansa = W ansa [ ] E ansa > W ansa [ ] W ansa > E ansa
Blue Filter (_____) *(check one):* [ ] E ansa = W ansa [ ] E ansa > W ansa [ ] W ansa > E ansa
Red Filter (_____) *(check one):* [ ] E ansa = W ansa [ ] E ansa > W ansa [ ] W ansa > E ansa

Note: Attach to this form all descriptions of morphology of atmospheric detail, as well as other supporting information. Please <u>do not</u> write on the back of this sheet. The intensity scale employed is the *Standard ALPO Intensity Scale*, where 0.0 = completely black ⇔10.0 = very brightest features, and intermediate values are assigned along the scale to account for observed intensity of features.

Association of Lunar and Planetary Observers (ALPO) Saturn Section: Visual Observation of Saturn for $B = \pm 18°$ to $\pm 20°$

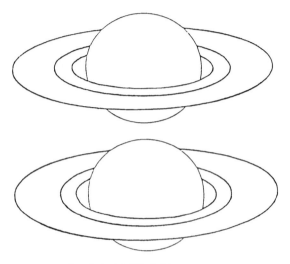

Coordinates *(check one)*: [ ] IAU [ ] Sky

Observer_____ Location_____

UT Date (start) _____ UT Start _____ CM I (start) _____° CM II (start) _____° CM III (start) _____°

UT Date (end) _____ UT End _____ CM I (end) _____° CM II (end) _____° CM III (end) _____°

B = _____° B = _____° Instrument _____ Magnification(s) _____ X_{min} _____ X_{max}

Filter(s) IL(none) _____ f_1 _____ f_2 _____ f_3 _____ Seeing _____ Transparency _____

Saturn Global and Ring Features	Visual Photometry and Colorimetry				Absolute Color Estimates	Latitude Estimates ratio y/r
	IL	f_1	f_2	f_3		

Bicolored Aspect of the Rings: No Filter (IL) *(check one)*: [ ] E ansa = W ansa [ ] E ansa > W ansa [ ] W ansa > E ansa
(always use IAU directions) Blue Filter (_____) *(check one)*: [ ] E ansa = W ansa [ ] E ansa > W ansa [ ] W ansa > E ansa
Red Filter (_____) *(check one)*: [ ] E ansa = W ansa [ ] E ansa > W ansa [ ] W ansa > E ansa

Note: Attach to this form all descriptions of morphology of atmospheric detail, as well as other supporting information. Please do not write on the back of this sheet. The intensity scale employed is the *Standard ALPO Intensity Scale*, where 0.0 = completely black ⟺10.0 = very brightest features, and intermediate values are assigned along the scale to account for observed intensity of features.

Association of Lunar and Planetary Observers (ALPO) Saturn Section: Visual Observation of Saturn for *B* = ± 22° to ± 24°

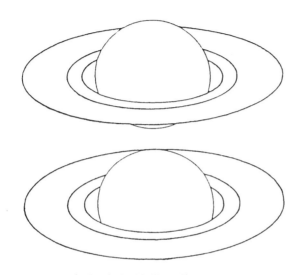

Coordinates (check one): [ ] IAU [ ] Sky

Observer_____ Location_____

UT Date (start) _____ UT Start _____ CM I (start) _____° CM II (start) _____° CM III (start) _____°

UT Date (end) _____ UT End _____ CM I (end) _____° CM II (end) _____° CM III (end) _____°

B = _____° B = _____° Instrument _____ Magnification(s) _____ x$_{min}$ _____ x$_{max}$

Filter(s) IL(none) _____ f$_1$ _____ f$_2$ _____ f$_3$ _____ Seeing _____ Transparency _____

Saturn Global and Ring Features	Visual Photometry and Colorimetry				Absolute Color Estimates	Latitude Estimates ratio y/r
	IL	f$_1$	f$_2$	f$_3$		

Bicolored Aspect of the Rings: No Filter (IL) *(check one)*:   [ ] E ansa = W ansa   [ ] E ansa > W ansa   [ ] W ansa > E ansa
(always use IAU directions)   Blue Filter (_____) *(check one)*:   [ ] E ansa = W ansa   [ ] E ansa > W ansa   [ ] W ansa > E ansa
         Red   Filter (_____) *(check one)*:   [ ] E ansa = W ansa   [ ] E ansa > W ansa   [ ] W ansa > E ansa

Note: Attach to this form all descriptions of morphology of atmospheric detail, as well as other supporting information. Please <u>do not</u> write on the back of this sheet. The intensity scale employed is the *Standard ALPO Intensity Scale*, where 0.0 = completely black ⟺ 10.0 = very brightest features, and intermediate values are assigned along the scale to account for observed intensity of features.

Association of Lunar and Planetary Observers (ALPO) Saturn Section: Visual Observation of Saturn for $B = \pm 26°$ to $\pm 28°$

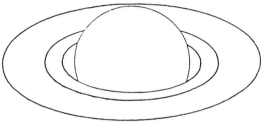

Coordinates (check one): [ ] IAU   [ ] Sky

Observer_____ Location_____

UT Date (start) _____ UT Start _____ CM I (start) _____° CM II (start) _____° CM III (start) _____°

UT Date (end) _____ UT End _____ CM I (end) _____° CM II (end) _____° CM III (end) _____°

B = _____ ° B = _____ ° Instrument _____ Magnification(s) _____ x_{min} _____ x_{max}

Filter(s) IL(none) _____ f_1 _____ f_2 _____ f_3 _____ Seeing _____ Transparency _____

Saturn Global and Ring Features	Visual Photometry and Colorimetry				Absolute Color Estimates	Latitude Estimates ratio y/r
	IL	f_1	f_2	f_3		

Bicolored Aspect of the Rings: No Filter   (IL) (*check one*):   [ ] E ansa = W ansa   [ ] E ansa > W ansa   [ ] W ansa > E ansa
(always use IAU directions)   Blue Filter   (_____) (*check one*):   [ ] E ansa = W ansa   [ ] E ansa > W ansa   [ ] W ansa > E ansa
Red   Filter   (_____) (*check one*):   [ ] E ansa = W ansa   [ ] E ansa > W ansa   [ ] W ansa > E ansa

Note: Attach to this form all descriptions of morphology of atmospheric detail, as well as other supporting information. Please do not write on the back of this sheet. The intensity scale employed is the *Standard ALPO Intensity Scale*, where 0.0 = completely black ⇔ 10.0 = very brightest features, and intermediate values are assigned along the scale to account for observed intensity of features.

Date: _____

Second Page

Observer: _____

(Attach to first page of observing form)

Saturn Global and Ring Features	Visual Photometry and Colorimetry				Absolute Color Estimates	Latitude Estimates ratio y/r
	IL	f_1	f_2	f_3		

Appendix B:
Bibliography

Abell, G.O., et al, *Exploration of the Universe*. Philadelphia: Saunders, 1987.

Adamoli, G. 1982, 'Some European Visual Observations of Saturn in 1981,' *J.A.L.P.O., 29*, (7–8): 143–145.

_____ 1983, 'Visual Observations of Saturn in the 1981–82 Apparition,' *J.A.L.P.O., 30*, (1–2): 11–16.

_____ 1984, 'Saturn in 1982–83: Some European Observations,' *J.A.L.P.O., 30*, (9–10): 177–180.

_____ 1986, 'Visual Observations of Saturn in 1984,' *J.A.L.P.O., 31*, (7–8): 164–167.

_____ 1988, 'Visual Observations of Saturn in 1986,' *J.A.L.P.O., 32*, (9–10): 207–210.

_____ 1990, 'Visual Observations of Saturn in 1988,' *J.A.L.P.O., 34*, (2): 59–60.

Aguirre, E.L. 2001, 'Astro Imaging with Digital Cameras,' *Sky and Telescope, 102*, (2): 128–134.

Alexander, A.F. O'D. *The Planet Saturn*. London: Faber and Faber, 1962.

Beatty, J.K., et al, eds. *The New Solar System*. Cambridge: Sky Publishing, 1982 (revised edition).

Benton, J.L., Jr. *A Handbook for Observing the Planet Saturn*. Savannah: Review Publishing, 1971.

_____ *Visual Observations of the Planet Saturn and Its Satellites: Theory and Methods*. Savannah: Review Publishing Co., 1975.

_____ 1971, 'Aims of the A.L.P.O. Saturn Section,' *J.A.L.P.O., 23*, (1–2): 10–11.

_____ 1972, 'The 1967–68 and 1968–69 Apparitions of Saturn,' *J.A.L.P.O., 23*, (3–4): 44–51.

_____ 1972, 'The 1970–71 Apparition of Saturn,' *J.A.L.P.O., 23*, (11–12): 115–222.

_____ 1972, 'The 1969–70 Apparition of Saturn,' *J.A.L.P.O., 24*, (1–2): 27–35.

_____ 1973, 'The 1971–72 Apparition of Saturn,' *J.A.L.P.O., 24*, (7–8): 139–147.

_____ 1974, 'The 1972–73 Apparition of Saturn,' *J.A.L.P.O., 25*, (3–4): 72–78.

_____ 1975, 'The 1973–74 Apparition of Saturn,' *J.A.L.P.O., 25*, (9–10): 183–191.

_____ 1976, 'The 1966–67 Apparition of Saturn,' *J.A.L.P.O., 25*, (11–12): 232–252.

_____ 1976, 'Visual and Photographic Observations of Saturn: the 1974–75 Apparition,' *J.A.L.P.O., 26*, (5–6): 94–100.

_____ 1976, 'Latitudes of Saturnian Features by Visual Methods,' *J.B.A.A., 86*, (5): 383–385.

_____ 1976, 'A Simultaneous Observing Program for the Planet Saturn: Some Preliminary Remarks,' *J.A.L.P.O., 26*, (7–8): 164–166.

_____ 1977, 'The 1975–76 Apparition of Saturn,' *J.A.L.P.O., 26*, (9–10): 173–184.

_____ 1978, 'The 1976–77 Apparition of Saturn,' *J.A.L.P.O., 27*, (7–8): 137–151.

_____ 1979, 'The 1977–78 Apparition of Saturn,' *J.A.L.P.O., 27*, (11–12): 246–255.

_____ 1979, 'The 1977–78 Apparition of Saturn (Conclusion),' *J.A.L.P.O., 28*, (1–2): 13–17.

_____ 1979, 'The 1978–79 Apparition of Saturn,' *J.A.L.P.O., 28*, (7–8): 140–150.

_____ 1979, 'Observational Notes Regarding the 1979–80 Apparition of Saturn: the Edgewise Presentation of the Rings,' *J.A.L.P.O., 28*, (1–2): 1–5.

_____ 1983, 'The 1979–80 Apparition of Saturn and Edgewise Presentation of the Ring System,' (part 1) *J.A.L.P.O., 29*, (11–12): 234–248.

_____ 1983, 'The 1979–80 Apparition of Saturn and Edgewise Presentation of the Ring System,' (Concluded) *J.A.L.P.O., 30*, (1–2): 26–33.

_____ 1983, 'The 1980–81 Apparition of Saturn,' *J.A.L.P.O., 30*, (3–4): 65–75.

_____ 1984, 'A Beginner's Guide to Visual Observations of Saturn,' *J.A.L.P.O., 30*, (5–6): 89–96.

_____ 1984, 'The 1981–82 Apparition of Saturn: Visual and Photographic Observations,' *J.A.L.P.O., 30*, (7–8): 133–141.

_____ 1986, 'The 1982–83 and 1983–84 Apparitions of the Planet Saturn: Visual and Photographic Observations,' *J.A.L.P.O., 31*, (7–8): 150–164.

_____ 1986, 'Southern Globe and Ring Features of Saturn: A Summary Analysis of Mean Visual Relative Numerical Intensity Estimates from 1966 through 1980,' *J.A.L.P.O., 31*, (9–10): 190–198.

_____ 1987, 'The 1982–83 and 1983–84 Apparitions of the Planet Saturn: Visual and Photographic Observations,' *J.A.L.P.O., 31*, (7–8): 150–164.

_____ 1987, 'The 1984–85 Apparition of Saturn: Visual and Photographic Observations,' *J.A.L.P.O., 32*, (1–2): 1–12.

_____ 1988, 'The 1985–86 Apparition of Saturn: Visual and Photographic Observations,' *J.A.L.P.O., 32*, (9–10): 197–207.

_____ 1989, 'The 1986–87 Apparition of Saturn: Visual and Photographic Observations,' *J.A.L.P.O., 33*, (7–9): 103–111.

_____ 1990, 'The 1987–88 Apparition of Saturn: Visual and Photographic Observations,' *J.A.L.P.O., 34*, (2): 49–59.

_____ 1990, 'The 1988–89 Apparition of Saturn: Visual and Photographic Observations,' *J.A.L.P.O., 34*, (4): 160–169.

_____ 1991, 'Monitoring Atmospheric Features on Saturn in 1991,' *J.A.L.P.O., 35*, (1): 24-25.

_____ 1991, 'A Photometric Opportunity for Saturn's Satellites,' *J.A.L.P.O., 35*, (2): 77.

_____ 1992, 'The 1990–91 Apparition of Saturn: Visual and Photographic Observations,' *J.A.L.P.O., 36*, (2): 49–62.

_____ 1993, 'Getting Started: Central Meridian (CM) Timings of Saturnian Atmospheric Features,' *J.A.L.P.O., 36*, (4): 155–156.

_____ 1993, 'The 1991–92 Apparition of Saturn: Visual and Photographic Observations,' *J.A.L.P.O., 37*, (1): 1–13.

_____ 1994, 'The 1992–93 Apparition of Saturn: Visual and Photographic Observations,' *J.A.L.P.O., 37*, (3): 97–107.

_____ 1995, 'The 1993–94 Apparition of Saturn: Visual and Other Observations,' *J.A.L.P.O., 38*, (3): 114–125.

_____ 1998, 'Observations of Saturn During the 1994–95 Apparition.' *J.A.L.P.O., 40*, (1): 1–13.

_____ 1999 'The 1995–96 Apparition of Saturn and Edgewise Presentation of the Rings: Visual and Other Observations,' *J.A.L.P.O., 41*, (1): 1–23.

_____ 2000, 'Observations of Saturn During the 1996–97 Apparition,' *J.A.L.P.O., 42*, (1): 1–12.

_____ 2001, 'Observations of Saturn During the 1998–99 Apparition,' *J.A.L.P.O., 43*, (4): 31–43.

_____ 2002, 'Observations of Saturn During the 1999–2000 Apparition,' *J.A.L.P.O., 44*, (1): 15–27.

_____ 2003, 'ALPO Observations of Saturn,' *Sky and Telescope, 106*, (6): 105.

_____ 2004, 'Saturn: A.L.P.O. Observations During the 2001–2002 Apparition,' *J.A.L.P.O., 46*, (1): 24–39.

_____ 2004, 'Saturn in Prime Time,' *Astronomy, 32*, (1): 88–92.

Blanco, V.M., and McCuskey, S.W. *Basic Physics of the Solar System.* Boston: Addison and Wesley, 1961.

Bobrov, M.S. *The Rings of Saturn.* Moscow: Nauka Press, 1970.

Borra, J.F., 1990, 'An Observer's Account of the Occultation of 28 Sagittarii By Saturn: 1989 July 03,' *J.A.L.P.O., 34*, (1): 22–23.

Budine, P.J. 1961, 'Amateur Observations of Saturn,' *J.A.L.P.O., 15*, (5–6): 80–82.

Burns, J., and Matthews, M.H., eds. *Satellites.* Tucson: University of Arizona Press, 1986.

Capen, C.F. 1958, 'Filter Techniques for Planetary Observers,' *Sky and Telescope, 17,* (12): 517–520.

_____, 1978, 'Recent Advances in Planetary Photography,' *J.A.L.P.O., 27,* (7–8): 47–51.

_____, and Parker, D.C. 1980, 'New Developments in Planetary Photography,' *J.A.L.P.O., 28,* (3–4): 45–50.

Chapman, C.R., and Cruikshank, D.P. *Observing the Moon, Planets, and Comets.* Unpublished manuscript.

_____ 1961, 'A Simultaneous Observing Program,' *J.A.L.P.O., 15,* (5–6): 90–94.

_____ 1962, 'The 1961 A.L.P.O. Simultaneous Observing Program – Second Report,' *J.A.L.P.O., 16,* (5–6): 134–140.

_____ 1963, 'A.L.P.O. Simultaneous Observing Program Schedule, September – November, 1963,' *J.A.L.P.O., 17,* (5–6): 112–113.

Cragg, T.A. 1961, 'Saturn in 1960,' *J.A.L.P.O., 15,* (7–8): 124–132.

Cragg, T.A., and Goodman, J.W. 1965, 'Saturn in 1963,' *J.A.L.P.O., 18,* (7–8): 132–140.

Cragg, T.A. 1966, 'Saturn's Edgewise Ring Presentation During 1966,' *J.A.L.P.O., 19,* (5–6): 73–75.

Cragg, T.A., and Bornhurst, L.A. 1966, 'A Preliminary Report Upon the 1965–66 Saturn Apparition,' *J.A.L.P.O., 19,* (9-10): 170–171.

Cragg, T.A., and Bornhurst, L.C. 1968, 'The 1965-66 Apparition of Saturn,' *J.A.L.P.O., 21,* (3–4): 54–60.

Davis, M. 2003, 'Shooting the Planets with Webcams,' *Sky and Telescope, 105,* (6): 117–122.

Delano, K.J. 1971, 'The Brightness of Iapetus,' *J.A.L.P.O., 22,* (11–12): 206.

Dollfus, A., ed. *Surfaces and Interiors of Planets and Satellites.* New York: Academic Press, 1970.

Eastman Kodak Co., 1966, 'Kodak Wratten Filters for Scientific and Technical Use,' *Pamphlet,* 22nd. ed. Rochester: Kodak, revised.

Elliot, J., and Kerr, R. *Rings.* Cambridge: MIT Press, 1984.

Gehrels, T., and Matthews, M.S., eds. *Saturn.* Tucson: University of Arizona Press, 1984.

Goodman, J.W., et al, 1962, 'An Occultation of BD 19°.5925 by Saturn and Its Rings on July 23, 1962: Observations Requested,' *J.A.L.P.O., 16,* (5–6): 131–133.

Goodman, J.W., et al, 1963, 'Saturn in 1962,' *J.A.L.P.O., 17,* (5–6): 91–97.

Gordon, R.W. 1979, 'Resolution and Contrast,' *J.A.L.P.O., 27,* (9–10): 180–189.

Grafton, E. 2003, 'Get Ultrasharp Planetary Images with Your CCD Camera,' *Sky and Telescope, 106,* (3): 125–128.

Guerin, P. 1970, 'A New Ring of Saturn,' *Sky and Telescope, 40,* (2): 88.

Haas, W.H. 1967, 'Latitudes on Saturn: A Note on Comparing Methods,' *J.A.L.P.O., 20,* (7–8): 133–135.

_____ 1981, 'Selected Drawings from the 1965–66 Edgewise Presentation of the Rings of Saturn,' *J.A.L.P.O., 28,* (11–12): 228–230.

_____ 1993, 'A Sample Study of the Rotation of the 1990 Equatoirial Zone Great White Spot on Saturn,' *J.A.L.P.O., 36,* (4): 151–153.

Harrington, P.S. *Starware,* 2nd ed. New York: John Wiley & Sons, 1998.

Hartmann, W.K. *Moons and Planets,* 4th ed. San Francisco: Wadsworth Publishing, 1995.

_____ 1975, 'Saturn – The New Frontier,' *Astronomy, 3,* (1): 26–34.

Heath, A.W. 1980, 'Some Recent Notes on Saturn,' *J.A.L.P.O., 28,* (7–8): 165–166.

Hodgson, R.G. 1970, 'Orbital Inclinations as a Factor in Satellite Light Variations,' *J.A.L.P.O., 22,* (1–2): 36.

Horne, J. 2003, 'Four Low-Cost Astronomical Video Cameras,' *Sky and Telescope, 105,* (2): 57–62.

Horne, J. 2001, 'The Astrovid Color Planetcam,' *Sky and Telescope, 102,* (2): 55–59.

Jet Propulsion Laboratory, 1977–1981, 'Voyager Bulletin,' *Mission Status Reports No. 1–61.* Pasadena: Jet Propulsion Laboratory (NASA).

Kingslake, R. *Optical System Design.* New York: Academic Press, 1983.

Kuiper, G.P., and Middlehurst, B.M., eds. *Planets and Satellites.* Chicago: University of Chicago Press, 1961.

Lavega, A.S. 1978, 'Nomenclature of Saturn's Belts and Zones (Southern Hemisphere),' translated by J.L. Benton, Jr. *J.A.L.P.O.*, *27*, (7–8): 151–154.

Le Grand, Y. *Light, Color, and Vision*. New York: Wiley, 1957.

McEwen, A.S. 2004, 'Journey to Saturn,' *Astronomy, 32*, (1): 34–41.

Mollise, R. *Choosing and Using a Schmidt–Cassegrain Telescope*. London: Springer-Verlag, 2001.

Morrison, D. *Voyage to Saturn*. Washington: U.S. Government Printing Office, 1982 (NASA SP-451).

Muirden, J. *The Amateur Astronomer's Handbook*. New York: Crowell, 1968.

Naeye, R. 2005, "News Notes: A Flood of Cassini Discoveries," *Sky and Telescope, 109*, (3): 16–17.

NASA, 1980–2005, *Planetary Photojournal* (Web site – various public domain images). Pasadena: Jet Propulsion Laboratory (California Institute of Technology).

Optical Society of America. 1963, *The Science of Color*. Ann Arbor, MI: Committee on Colorimetry of the Optical Society of America, 1963.

de Pater, I., and Lissauer, J.J. *Planetary Sciences*. New York: Cambridge University Press, 2001.

Paul, H.E. *Telescopes for Skygazing*. New York: Amphoto, 1976, revised.

Peach, D.A. 2003, 'Saturn at Its Most Spectacular,' *Sky and Telescope, 106*, (6): 103–107.

Peek, B.M. *The Planet Jupiter*. London: Faber and Faber, 1968.

Proctor, R.A. *Saturn and Its System*. London: Longmans, 1865.

Reese, E.J. *Measurements of Saturn in 1969*. Las Cruces: New Mexico State University Observatory Publications, 1970.

Roth, G.D. *Handbook for Planetary Observers*. London: Faber and Faber, 1971.

Rotherty, D.A. *Satellites of the Outer Planets*, 2nd ed. New York: Oxford University Press, 1999.

Rutten, H., and van Venrooij, M. *Telescope Optics: Evaluation and Design*. Richmond: Willmann-Bell, 1988.

_____ *Tashenbuch für Planetenbeobachter*. Mannheim: Bibliographisches Institut, 1966.

Sassone-Corsi, E., and Sassone-Corsi, P. 1966, Some Systematic Observations of Saturn During Its 1974–75 Apparition, *J.A.L.P.O., 26*, (1–2): 8–12.

_____ 1979, 'Italian Observations of Saturn During 1975–78,' *J.A.L.P.O., 27*, (11–12): 222–225.

_____ 1981, 'Hypothetical Spectral Variability of Titan,' *J.A.L.P.O., 28*, (11–12): 230–234.

_____ 1981, 'New Statistical Measurements of Saturn's Rings,' *J.A.L.P.O., 29*, (1–2): 24–27.

Schmidt, I. 1960, 'The Green Areas of Mars and Colour Vision,' *Proceeding of the 10th Annual International Astronautical Congress*, 171–180.

Sharonov, V.V. *The Nature of the Planets*. Jerusalem: IPST, 1964.

Sheehan, W. 1980, 'On an Observation of Saturn: The Eye and the Astronomical Observer,' *J.A.L.P.O., 28*, (7–8): 150–154.

Sidgwick, J.B. *Amateur Astronomer's Handbook*, 3rd ed. London: Faber and Faber, 1971.

_____ *Observational Astronomy for Amateurs*, 3rd ed. London: Faber and Faber, 1971.

Sitler, J. 1963, 'The Origin and Development of the Dollfus White Spot on Saturn,' *J.A.L.P.O., 16*, (11–12): 251–253.

Slipher, E.C. *A Photographic Survey of the Brighter Planets*. Cambridge: Sky Publishing, 1964.

Smith, E., and Jacobs, K. *Introductory Astronomy and Astrophysics*. Philadelphia: Saunders, 1973.

Spilker, L.J., ed. *Passage to a Ringed World: The Cassini-Huygens Mission to Saturn and Titan*. Washington, DC: U.S. Government Printing Office, 1997 (NASA SP-533).

Stevens, S.S., ed. *Handbook of Experimental Psychology*. New York: Wiley, 1951.

Suiter, H.R. *Star Testing Astronomical Telescopes*. Richmond: Willmann-Bell, 1994.

Taylor, S.R. *Solar System Evolution – A New Perspective*. 2nd ed. New York: Cambridge University Press, 2001.

Tytell, D. 2004, 'NASA's Ringmaster,' *Sky and Telescope, 108*, (5): 38–42.

_____ 2005, 'Titan: A Whole New World,' *Sky and Telescope, 109*, (4): 34–38.

Van de Hulst, H.C. *Light Scattering By Small Particles*. New York: Wiley, 1957.

de, Vaucouleurs, G. *Physics of the Planet Mars*. London: Faber and Faber, 1954.

Vitous, J.P. 1962, 'Observations of Planetary Color,' *J.A.L.P.O., 16*, (1–2): 35–37.

Westfall, J.E. 1970, 'Saturn Central Meridian Ephemeris, January, 1970 – December, 1970,' *J.A.L.P.O., 22*, (1–2): 36.

_____ 1979, 'Selected Phenomena of Saturn's Satellites: The Fall, 1979, Ring Passage,' *J.A.L.P.O., 28*, (1–2): 5-13.

_____ 1980, 'Mutual Phenomena of Saturn's Satellites Titan and Rhea: 1980, January – September 22,' *J.A.L.P.O., 28*, (3–4): 55–61.

_____ 1980, 'Mutual Phenomena of Saturn's Brighter Satellites: July – August, 1980,' *J.A.L.P.O., 28*, (5–6): 112–116.

_____ 1980, 'Recent Observations of Saturn, With Prospects for Spring and Summer, 1980,' *J.A.L.P.O., 28*, (5–6): 124–125.

_____ 1980, 'Some Observations of Saturn During the Current 1979-80 Apparition,' *J.A.L.P.O., 28*, (9–10): 184–190.

_____ 1984, 'Saturn Central Meridian Ephemeris: 1985,' *J.A.L.P.O., 30*, (11–12): 247–249.

Whipple, F.L. *Earth, Moon, and Planets*, 3nd ed. Cambridge: Harvard University Press, 1969.

Woodworth, R.S., and Schlosberg, H., *Experimental Psychology*. New York: Holt, 1954 revised.

Index

A

AAVSO star charts, 142
Airy disk, 77
ALPO
—eGroup, 3
—forms, 164–173
 —central meridian (CM) transit
  data, 164
 —sectional sketches, 164
 —visual observation of Saturn
  (by *B* value), 166–173
 —visual observation of Saturn's
  satellites, 165
—observing programs, 3, 112
—Relative Numerical Intensity
 Scale, 123
—*Saturn CM longitudes*, 137
—Saturn Relative Numerical
 Intensity
 —Scale, 123, 124, 126
—Saturn Section, 2
—seeing scale, 76
—system I, II, and III data, 111,
 137
—Web site, 3
amateur astronomers, endeavors
 involved in, 90
Antoniadi seeing scale, 76
aperture, effective, 79, 80
Association of Lunar and Planetary
 Observers *see* ALPO
asteroids, 7
Astronomical Almanac, ephemeris
 data, 111, 137
astronomical seeing, 75–82
—ALPO seeing scale, 75
—Antoniadi seeing scale, 75
astrovideography, 148–151
Atlas, 23, 49
atmosphere of Saturn, 13–19
—auroras, 17, 19
—cloud layers, 15–16, 162–163
—composition, 13
—ionosphere, 15
—polar vortices, 14
—scattering of light by, 160–163
—stratosphere, 15
—tropopause, 15
—vertical structure, 14
—white spots, 16, 18, 95–97, 158
—wind patterns, 16–17, 18
auroras, 17, 19

B

B (Saturnicentric latitude of
 Earth), 89, 113, 134
B´ (Saturnicentric latitude of sun),
 134
BAA
—intensity scale, 126
—observing programs, 3, 112
—system I, II, and III data, 111,
 137
Barlow lenses, 68–69
Bezold-Brüke phenomenon, 88
British Astronomical Association
 see BAA

C

Calypso, 49
camcorders, 148
Cassini Regio, 47
Cassini's division (A0/B10), 12,
 22–23, 27, 29, 83
—basic data, 24
—telescopic appearance, 99, 102
catadioptrics, 57–59
—for colorimetry, 128
—costs, 59
—focal ratios, 59
—Maksutov-Cassegrain (MAK),
 57–59, 61
—portability, 59
—Schmidt-Cassegrain (SCT),
 57–59, 61–62
—spherical aberration, 58
CCD cameras, 151–152
CCD chips, sensitivity to infrared
 wavelengths, 131, 158
central meridian (CM)
—transit timings, 136–140
 —ALPO form, 164
Charon, 7
Circus Maximus, 39
color
—contrast-induced, 88
—standard abbreviations, 130
color filters, 70–72, 73
—techniques for use, 126–129
—"universal", 128
—Wratten, 71–72, 73, 127, 128
 —properties, 73
 —recommended tricolor series
  by aperture, 128

color filters *(continued)*
 —usage for Saturn observations,
  73
color perception, 87–88
color reference charts, 88
color sensitivity, 121, 127
comets, 7
cones, 87, 127
contrast, 85–87
—apparent, 85
—true, 85
contrast perception, 86–87
contrast sensitivity, 86, 121
correlation coefficient, personal,
 82
Crape ring (ring C), 12, 24, 25–26
—at edgewise apparitions, 105
—ringlets, 26
—telescopic appearance, 99, 101

D

Dawes's limit, 83
—comparison with Rayleigh's
 criterion, 84
declination (DEC), 60
Deimos, 5
"density wakes", 30
digital cameras, 152–155
—zoom function, 153
Dione, 33, 45
—bulk density, 45
—co-orbital satellite, 49
—craters, 45
—observation, 141, 142
double stars, for estimates of
 seeing, 81
drawing Saturn's globe and rings,
 111–122
—blanks for use *see* ALPO, forms
—equipment, 113
—executing drawing, 113–118
—factors affecting reliability,
 120–122
—field orientation, 118–120
—nomenclature, 118
—objective narrative, 116–118
—purposes and objectives,
 111–113
—sectional sketches, 115, 116
 —ALPO form, 164
—shadows, 118–119

drawing Saturn's globe and rings
   (continued)
—start and end time recording,
   115
—strip sketches, 115
—supporting data, 115–116, 117

E
Earth, 5
—magnetosphere, 20–21
eccentric (mean) latitude (*E*), 134,
   135
efficiency, of telescope, 79–80
electronic imaging eyepieces, 150
Enceladus, 33, 41–44
—bulk density, 41
—craters, 43–44
—H_2O volcanism, 43, 44
—observation, 141, 142
—reflectivity, 43
Encke's complex/division (E5),
   12–13, 24, 30–31
—telescopic appearance, 99, 102
Epimetheus, 23, 33, 49
equatorial zone (EZ), 10, 95–97,
   105, 163
—EB (equatorial belt), 10, 97
—EZn, 10, 95
—EZs, 10, 95
exit pupil, 66
eye
—insensitivity to UV light, 161
—photo-receptive cells, 87
—sensitivity to IR wavelengths,
   161–162
eye relief, 66
eyepieces, 64–68
—electronic imaging, 150
—field of view
      —apparent, 65
      —true, 65–66
focal lengths, 65
parfocal, 65, 68
types, 67–68

F
field orientation, 118–120
field of view
—apparent, 65
—true, 65–66
filter techniques, 126–129
finder telescopes, 69–70
flip mirror assembly, 157–158
focal length
—eyepiece, 65
—telescope, 65
fork equatorial mountings,
   61–62

G
Ganymede, 36
German equatorial mountings,
   60–61
"go-to" mounts, 60
Great White Spot (1990), 95–96

Guerin gap, 24, 25

H
Haas, Walter H., 2, 135
Haas technique, 135
He precipitation, 19
Helene, 49
Herschel, 40, 42
Huygen's gap, 24, 29
Hyperion, 33, 48–49
—bulk density, 48
—craters, 48
—observation, 141, 142
—orbital resonance with Titan,
   48

I
Iapetus, 33, 46–47
—bulk density, 46
—craters, 46–47
—observation, 141, 142
IAU, convention on planetary
   observations, 9, 63, 120
imaging Saturn and its ring
   system, 147–163
—astrovideography, 148–151
—CCD imaging, 151–155
—systematic imaging, 159–163
—webcams, 155–157
      —image capturing and
         processing procedure,
         157–158
International Astronomical Union
   see IAU
irradiation, 127
Ithaca Chasma, 44, 45

J
Janus, 23, 33, 49
Johnson UBV filter sets, 132
Jupiter, 5
—aurora displays, 163
—cloud layers, 15–16
—CM longitude data, 137
—interior, 19, 20
—internal heat radiation, 163
—irregular satellites, 49
—magnetosphere, 20–21
—rings, 11
—rotation, 136
—wind patterns, 17

K
Keeler's gap (A8), 13, 24, 31, 100
—telescopic appearance, 102
Kellner eyepieces, 67, 68
Kuiper belt, 7

L
latitude measurement and
   estimation, 133–136
—on CCD and webcam images,
   133–134

latitude measurement and
   estimation *(continued)*
—from drawings, 133
—Haas technique, 135
—on high-resolution photographs,
   133
—using filar micrometer, 134
luminance, 83–84

M
magnification, 65–68
—maximum, 67
—minimum useful, 66–67
Maksutov–Cassegrain telescope
   (MAK), 57–59, 61
Mars, 5
Maxwell division, 24, 26
Mercury, 5
—phase angle, 84
—visual geometric albedo, 84
mesopic vision, 87
meteorological conditions, optimal
   for seeing, 80
Mimas, 23, 29, 33, 40–41
—bulk density, 40
—craters, 33, 40–41, 42
—observation, 141–142
moon
—occultations of Saturn by,
   108–110
—phase angle, 84
—size, 3
moonlets, 22, 23

N
Neptune, 5, 11, 163
north equatorial belt (NEB), 10,
   97–98
—NEBn, 10, 97
—NEBs, 10, 97
—NEBZ (north equatorial belt
   zone), 10, 97
north north temperate belt
   (NNTeB), 10, 98
north north temperate zone
   (NNTeZ), 10, 98
north polar belt (NPB), 99
north polar cap (NPC), 11, 99
north polar region (NPR), 11, 99
north temperate belt (NTeB), 10, 98
north temperate zone (NTeZ), 10,
   98
north tropical zone (NTrZ), 10, 98

O
observations
—ALPO forms for recording *see*
   ALPO, forms
—simultaneous, 88
—systematic, 75
occultations
—of Saturn by moon, 108–110
—of stars by Saturn's globe and
   rings, 107–108
oculars *see* eyepieces

Odysseus, 44–45
Oört cloud, 7
orthoscopic eyepieces, 67–68

P
Pan, 22, 49
Pandora, 32, 49
phase angle, 84
Phobos, 5
Phoebe, 33, 49–50
—craters, 50
photometers
—CCD, 132
—photoelectric, 132, 143
photopic vision, 87, 127
planetary video cameras, 149
planetocentric latitude, 134, 135
planetographic latitude (G), 134–135
planets
—inferior, 5
  —phase angles, 84
—Jovian, 5–7
—superior, 5
  —phase angles, 84
—terrestrial, 5
Plössl eyepieces, 68
Pluto, 7
polar axis, 60
polar vortices, 14
polarizers, variable-density, 72, 74
Prometheus, 32, 49
pupillary diameter, 66, 78
Purkinje effect, 87, 127

R
Rayleigh's criterion, 78, 83, 87
—comparison with Dawes's limit,
  84
reflectors, 55–57
—Cassegrain, 52, 55–57
—for colorimetry, 128
—focal ratios, 55
—maintenance, 55–56
—Newtonian, 52, 54, 55–57
—portability, 55
reflex sights, 69–70, 71
refractors, 51–55
—achromatic, 52, 53–55, 128
—advantages, 51
—apochromatic, 55
—chromatic aberration, 53, 54–55
—costs, 53, 55
—focal ratios, 53–54
—portability, 53, 55
resolution
—of eye, minimum, 78
—theoretical limits, 84
Rhea, 33, 45–46
—bulk density, 45
—craters, 33, 46
—observation, 141, 142
right ascension (RA), 60
ring system of Saturn, 8, 11–13,
  21–33
—basic data, 24

ring system of Saturn *(continued)*
—bicolored aspect, 102–103
  —investigation, 130–132
—detailed view, 24
—drawing *see* drawing Saturn's
  globe and rings
—edgewise orientations
  —appearance, 103–107
  —dates, 104
—extraplanar ring particles, 107
—imaging *see* imaging Saturn and
  its ring system
—intensity minima, 99–100, 113
  —denotation, 100, 118
—mass, 21
—origin, 21
—outer edge sharpness, 23
—ring A, 12, 24, 30–31
  —at edgewise apparitions, 105
  —azimuthal brightness
    asymmetries, 30, 130–132
  —ringlets, 30, 31
  —telescopic appearance, 99, 102
—ring B, 12, 24, 27–29
  —at edgewise apparitions, 105
  —radial spokes, 28–29
  —ringlets, 27
  —telescopic appearance, 99,
    101–102
—ring C, 12, 24, 25–26
  —at edgewise apparitions, 105
  —ringlets, 26
  —telescopic appearance, 99, 101
—ring D, 13, 24, 25, 99
—ring E, 13, 24, 33, 99
  —at edgewise apparitions, 105
—ring F, 13, 24, 31–32, 99
—ring G, 13, 24, 33, 99
—telescopic appearance, 99–103
—tilt, 11–12
rods, 87, 127

S
S/2004 S1, 49
S/2004 S2, 49
satellites of Saturn, 33–50
—at edgewise presentations of
  rings, 105–106
—basic data, 34–35
—eclipses, 145
—finder charts, 143
—irregular, 49–50
—magnitude estimation, 142–143
—montage, 36
—observation, 141–146
  —ALPO form, 165
—occultations, 144–145
—shadow transits, 144–145
—shepherding, 32, 33, 49
—transits, 144–145
—*see also individual satellites*
Saturn
—atmosphere *see* atmosphere of
  Saturn
—aurora displays, 163
—basic characteristics, 7–13

Saturn *(continued)*
—belts, 9
  —convective structure, 16
  —latitude determination *see*
    latitude measurement and
    estimation
  —nomenclature and
    characteristics, 10–11
—Bond albedo, 7
—diameter, 7
—drawing *see* drawing Saturn's
  globe and rings
—imaging *see* imaging Saturn and
  its ring system
—interior, 19–21
—internal heat radiation, 163
—magnetosphere, 20–21
—mass, 11
—mean density, 11, 13
—minimum useful magnification,
  66
—oblateness, 8
—obliquity, 8, 89
—perihelion, 94
—phase angle, 84
—ring system *see* ring system of
  Saturn
—Roche limit, 12, 22
—satellites *see* satellites of Saturn
—sidereal rotation periods, 11,
  111
—surface brightness, 84
  —apparent, 85
—synodic period, 7, 89
—system I *see* system I
—system II *see* system II
—system III *see* system III
—telescopic appearance of globe,
  90–99
  —equatorial zone, 95–97
  —northern hemisphere, 97–99
  —southern hemisphere, 91–95
—thermal response to solar
  heating, 94
—visual geometric albedo, 8, 84
—zones, 9
  —convective structure, 16
  —nomenclature and
    characteristics, 10–11
Saturnicentric latitudes
—of Earth (B), 89, 113, 134
—of feature (C), 134, 135
—of sun (B´), 134
Saturnigraphic latitude, of feature
  (G), 134–135
Schmidt-Cassegrain telescope
  (SCT), 57–59, 61–62
scotopic vision, 87
sectional sketches, 115, 116
—ALPO form, 164
security cameras, 149
seeing *see* astronomical seeing
shadows
—of globe on rings (Sh G on R),
  107
—representation on drawings,
  118–119

shadows *(continued)*
—of rings on globe (Sh R on G), 107
solar heating, 94
solar system, simplified view, 5–7
south equatorial belt (SEB), 10, 92–94
—SEBn, 10, 92, 93
—SEBs, 10, 92, 93
—SEBZ (south equatorial belt zone), 10, 92
south polar belt (SPB), 91
south polar cap (SPC), 10, 91
south polar region (SPR), 10, 91
south south temperate belt (SSTeB), 10, 92
south south temperate zone (SSTeZ), 10, 91
south temperate belt (STeB), 10, 92
south temperate zone (STeZ), 10, 92
south tropical zone (STrZ), 10, 92, 93
stacking, 158
star charts, AAVSO, 142
star diagonals, 63–64, 119
strip sketches, 115
Strolling Astronomer, The, 2
sun, composition, 20
surface brightness, 84
—apparent, 85, 86
system I, 11, 111, 136, 137
—motion of CM longitude in intervals of mean time, 138
—rotation rate, 11, 111, 136–138
system II, 11, 111, 136, 137
—motion of CM longitude in intervals of mean time, 138
—rotation rate, 11, 111, 136–138
system III, 11, 111, 136
—motion of CM longitude in intervals of mean time, 138
—radio rate, 11, 111, 136–138

T
telescopes, 51–74
—accessories, 63–74
—categories, 51
—choosing, for observing Saturn, 74
—efficiency, 79–80
—finder, 69–70
—focal length, 65
—mountings, 59–62
—computer-controlled, 60
—equatorial, 59–60
—*see also* catadioptrics; reflectors; refractors
Telesto, 49
Terby white spot (TWS), 107, 108
Tethys, 33, 44–45
—bulk density, 44
—co-orbital satellites, 49
—craters, 44–45
—observation, 141, 142
Titan, 33, 36–40
—atmosphere, 6, 33, 36–37
—bulk density, 39
—CH_4 absorption filter imaging, 145
—data from UV and IR regions, 160
—ground fog, 38, 42
—haze layer, 36, 38
—hue, 36, 37, 38, 41
—infrared imaging, 145
—interior structure, 39
—and magnetosphere, 21
—observation, 141, 142, 145–146
—orbital resonance with Hyperion, 48
—"petrochemical rain", 37
—"seas", 38
—surface material, 37, 39
topography, in astronomical seeing quality, 80–82
transparency, atmospheric, 82–83

U
ultraviolet (UV) light, 160–163
Universal time (UT), 138
Uranus, 5, 11

V
Venus, 5
—phase angle, 84
—visual geometric albedo, 84
video cameras, planetary, 149
video capture cards, 150
vignetting, 153, 154
visual acuity, 120
visual color estimates, absolute, 129–130
visual colorimetry, 126–129
visual numerical relative intensity scales
—ALPO, 123, 124, 126
—BAA, 126
visual photometry, 123–126
—of satellites, 142–143
"visual purple", 87

W
webcams, 155–157
—image capturing and processing procedure, 157–158
white spots, 16, 18, 95–97, 158
Wratten color filters, 71–72, 73, 127, 128
—properties, 73
—recommended tricolor series by aperture, 128
—usage for Saturn observations, 73

X
Xanadu, 38, 40